Firecall

True Stories of Irish Firefighting and Rescue

Ruairi Kavanagh

Gill & Macmillan

Gill & Macmillan Ltd
Hume Avenue, Park West, Dublin 12
with associated companies throughout the world
www.gillmacmillan.ie

© Ruairi Kavanagh 2006
ISBN-13: 978 07171 3948 4
ISBN-10: 0 7171 3948 4

Typography design: Make Communication
Print origination: TypeIT, Dublin
Printed and bound by ColourBooks Ltd, Dublin

This book is typeset in Linotype Minion and Neue Helvetica.

The paper used in this book comes from the wood pulp of
managed forests. For every tree felled, at least one tree is
planted, thereby renewing natural resources.

A CIP catalogue record for this book is available from the British
Library.

5 4 3 2 1

Contents

Acknowledgments

This book is not a technical guide to fire and rescue, neither is it an authoritative guide to Irish fire services. Unlike other state agencies such as An Garda Síochána, there is no facility which enables the public to appreciate the work done and the risks taken by Irish firefighters. They wouldn't want one either, that's not their nature or part of their agenda. What I have tried to do is provide a snapshot in these pages of what it is like to be a fire and rescue professional, to face fire, danger, death and human misery as part of your regular day's work. There is no centralised reporting system for all the incidents which firefighters respond to, so my stories are mainly based on the experiences of firefighters and fire officers in first-hand interviews or else from press reports or other data which they may have had to hand. Some stories are anecdotal, some are historical and will be familiar to many people, and in these cases I have tried to add the recollections of firefighters to give a fresh angle to them where possible. I have also used some previous information which I sourced during my time as editor of *Firecall* magazine. All fire authorities in the country were contacted in the preparation of this book.

As I worked with Dublin Fire Brigade in a publishing capacity for many years, a large amount of the material here is based on the experiences of firefighters in the capital. That is no slight on other services, rather the basic fact that the vast and overwhelming amount of emergency calls stem from Dublin and represent a greater section of the fire service's work in general.

Fire and rescue does by its very nature contain stories of terrible tragedy, but I have tried not to focus heavily on these. The stories do of course refer to actual incidents but specifics, other than those already in the public domain, have been removed to protect the identities of those who may have been involved or affected. I have tried to be as technically accurate as possible in all cases.

There are many people to thank of course: Fergal Tobin at Gill & Macmillan for commissioning the book, and his excellent staff. I would also like to thank Ashville Media for their help with imagery and material, particularly Mark Mitchell and Andrew Dennehy, with whom I worked on *Firecall* magazine for several years. I would also like to thank my current colleagues and friends, in particular Máirtín Breathnach, Brian Clark and Michael Whelan for their patience, understanding and dark humour.

Firefighters are obviously the lifeblood of this book and there are so many to thank. Dublin Fire Brigade in particular have simply been phenomenally helpful. In particular, Station Officer Greg O'Dwyer has offered tireless assistance. Greg, in addition to giving me stories himself, was always a source of ideas for other stories in the book. I would also like to thank all the guys at D Watch in Finglas including Station Officer Paddy Hughes and the recently retired Jim 'Swill' Macdonald who, to me, epitomises the spirit of the brigade. Also very helpful in the brigade (serving and retired) were Alan Doyle, Eddie Reilly and Paul Carolan for taking such an interest along with Bob Murray, Ray Murray, Jim McDonald, Colin Murphy, Dave Connolly, Kevin Monaghan, Gerry Sterio, John Lynch, George Harrison, Joe O'Brien, Dave Kavanagh, Paul McGurrell, Damien Guilfoyle, Keith Carolan, Shay Rowe, Terry O'Connor, Gary Hayden, Tommy Ellis, Bertie Horgan, Joe Broughan, Peter Charles, Keith Carolan, Alan Finn and his excellent website *www.irishfireservices.com*, Nobby Clarke, Dermot Ryan, Gerry Bell, Tony Daly, John Small, Ray Leavy and anyone else who made me a cup of coffee, led me up a blind alley with a dodgy story or didn't kick me out of the station every time I asked a stupid question. Special thanks must also go to Tom Geraghty, former firefighter and renowned brigade historian whose excellent book on the brigade was an invaluable reference. In Carlow, Paul Curran was a great help as always. I would like to thank his s/o John Comerford and firefighters Pat Craddock, Mick Gahan, Darrell Hayden and Pat Bolger. In Roscommon, cfo Cathal McConn, Seamus Cafferty and Fergus Frain for their time and knowledge. In Mayo, cfo Seamus Murphy and all the lads of Belmullet station particularly Chris Murphy and Joe Ruddy. In Donegal, Eamonn Dowdican, PJ Clancy, Johnny Mulreany and all at their station in Bundoran who recalled the painful memories of the Central Hotel fire with such clarity. In Kilkenny, thanks to acfo John Collins and former s/o Joe Traynor. In Dundalk, special thanks to Jim Kerley.

It is impossible to mention everybody so I would like to thank anyone who took time to help me. On a personal level, I would like to thank my family, particularly my mother and father, and my wife, Virginia Tey, who has been such an inspirational support through everything. This book is dedicated to her and to firefighters everywhere.

Abbreviations and Glossary

Some firefighting abbreviations and terminology used in this book:

ACFO: Assistant Chief Fire Officer.

Appliance: Standard fire engine.

BA: Breathing Apparatus, equipment used by firefighters in toxic atmosphere.

Backdraught: A fire explosion when oxygen is introduced into an oxygen-depleted fire.

Callout/Turnout: Responding to an emergency call.

CFO: Chief Fire Officer, in overall command of a fire service in a county.

DFB: Dublin Fire Brigade

D/O: District Officer, in charge of stations in one district. Not applicable to all counties.

DSU: Distress Signal Unit, also known in some brigades as DSX, an automatic device which triggers an alarm should a firefighter remain prone or stationary for more then thirty seconds.

EMT: Emergency Medical Technician.

FF: Standard time firefighting personnel, whether full-time or retained (i.e. volunteers).

Flashover: When superheated gases on a fireground begin to ignite.

Knock Down: To knock down or extinguish a fire.

Making Down: To deploy a piece of firefighting or rescue equipment.

MER: Marine Emergency Response.

PPE: Personal Protective Equipment

RTA: Road Traffic Accident.

Snorkel: Hydraulic firefighting platform.

S/O: Station Officer, in command of fire station.

Special Appliance: Specialist rescue vehicle, such as rescue tender, turntable ladder or decontamination unit.

SRT: Swiftwater Rescue Technique: A set of procedures used for water safety training and water rescue.

Sub/Off: Sub Officer, second in command at a fire station.

One Night on the Front Line

This chapter is an amalgamation of stories gathered anecdotally from Dublin Fire Brigade. Names and locations have been changed throughout.

Saturday 9.10 pm Phibsborough Fire Station, Dublin Fire Brigade, Callsign, Number 5

'We had just finished parade when the first bell went off, the usual, never get a minute at the weekends. It was raining bad, lashing down, never makes the job easier. The s/o had just done the duty roster, I was on BA, Breathing Apparatus, for the night. Don't mind it as the night normally flies in since you're kept so busy. When you're on BA, you know that you're the one who's going to be going up against the fire first. Anyway, the bell went, and we moved quickly like we always have to in this job. Locker doors left half open, coffee cups still steaming half empty, toast left in the toaster. We were gone.

'Strapped in at the rear of the appliance we knew we were going to fight a fire. Our equipment was there behind us, ready for action. You learn to trust and respect your gear, that's why we spend so much time checking it and making sure it's right. It's what keeps you alive. The streets weren't too busy but with heavy rain the going was slower then normal, even with the blue lights. Inside the appliance there is always slagging and banter, part of the job.

'I remember this callout, seeing the blue lights reflected in the glistening blackness of the rain-soaked streets. The radio crackled with life in the driver's cab, more units were being dispatched. Not just from Number 5, our station, but from North Strand and HQ as well. A District Call. This could be a bad one, but some firefighters always think that on every firecall. That's the way a fire appliance is, you have the joker, the chirpy one, the quiet one, the moody one etc. The going was slow when we entered the estate, trying to find our way to the source of the call. For such a large fire it was hard to spot, I smelt the smoke before I saw the fire. Then I saw it.

'We were the second appliance on the scene. The next couple of minutes were a blur, I don't really remember them because they're the minutes you're on auto-pilot, getting out of the truck, talking to the officers, sizing up the fire and hoping the gardaí would show up soon and keep the locals at bay. Quite a large crowd had gathered and that was no surprise. Number 27 was ablaze, smoke billowing from the charred eaves, not much flame, but as most firefighters will tell you, that's not necessarily a good thing. Of course the real problem was that there were two people still inside. I remember looking at the s/o, who looked at the fire and arched his eyebrows. We were going in.

'Putting on a BA set is always unpleasant; you're basically placing most of your senses within a tight mask of rubber and plastic. When you switch the valve for the air tank on, it makes it a bit easier, but the visibility through the visor is never great and as your breathing rate increases the overall feeling is one of anxiety. You control it because you're used to it.

'The door of the house was partially open, blackened with smoke. Four of us were going in, there were two children inside, get them and get out. The inside of the house was destroyed with smoke; the flames had done most of their work but you could still see flame licking at the walls and timber of the staircase. We headed for the staircase and the other guys peeled off to search the downstairs rooms. The carpet was scorched off but the wood seemed solid enough, you learn to depend a lot on your sense of touch because your other senses are so retarded. Inside a BA mask, surrounded by smoke you can't smell much, your hearing is muffled and in a bad fire like this you can hardly see a thing.

'Higher up in the house I could hear the water hitting the flames and the constant humming of the vehicles outside. My feet moved up the steps carefully, my hands in front of me, touching carefully. On the seventh step, I tried to move a piece of debris out of my way, then I realised it wasn't debris. I turned to the guy behind me, gesturing. I bent down and turned the child over, no more then seven, her eyes closed and her skin blackened. It didn't look good. I passed her to

Sean, and he moved toward the door. We could do nothing for her here, the ambulances were waiting outside. Seeing children in that condition, dead or almost dead, is tough but you put it to the back of your mind, just keep moving.

'I was waiting for Sean to come back but I wanted to get to the upstairs rooms, the smoke was denser here, I couldn't see any flame except the odd flicker on what was left of the wallpaper. I heard Sean coming up the stairs behind me, I moved on. The bathroom and the master bedroom were empty; we shone our torches on nothing except the remains of a ruined home. We pulsed water on the floor and walls but there was precious little left to save. The door to the children's bedroom was half open, the opening thick with billowing smoke. We moved forward, I could feel the heat through the visor, intense and rising. Sean tapped me on the shoulder, reminding me to keep my head down. We went to our hands and knees, the air was cooler there but I could feel the searing heat across my back. The room wasn't big but we had no sense of perspective since we literally couldn't see anything. I moved forward, my back was perspiring heavily. In the corner of the room was a small fountain of flame amidst the remains of what had once been a television set. The seat, or source, of the fire. Sean pulsed short bursts of water into the air above us: this room was extremely hot now and was in danger of becoming a flashover — when the gases ignite and flames roll through the air.

'Beside the bed I felt something, something small, a child probably. I moved carefully to gather her up, the torch didn't tell me much, from what the fire and heat had done I wasn't sure if it was a girl or a boy, probably a little girl. I signalled to Sean and we both began to move back towards the door, he kept pulsing water into the cloud of fire gases above us, trying to lower the temperature. I moved deliberately towards the stairs, there was no need to run. Whether I went fast or slow would make no difference to this child. At the bottom of the stairs I saw the two guys who were searching the downstairs, they saw what I was holding and went out ahead of me. I waited at the doorway, shrouded in smoke.

'There was a crowd of people at the gate, straining to see. I stopped and turned my back, shielding what remained of the child who had died in her bed. I heard the rattle of a stretcher on the garden path. I turned and one of the lads passed me a blanket. I gently placed the small bundle on the stretcher and pulled the blanket over her. There was no need for anyone to see what was under the blanket. There were blue lights everywhere now, gardaí and our own lads, helping to control the crowd. I saw Sean exit the house behind me as arcs of water flowed over us on top of the flames which were now beginning to flicker through cracked tiles on the roof. I stopped and looked at the fire for a moment then turned around, taking off my BA mask. The cold night air and cool rain was refreshing. Someone said something to me and maybe someone else clapped me on the shoulder, I don't really remember. I got back in the appliance and we got ready to leave. The ambulances were gone. I didn't know if the first girl made it or not, I already knew the fate of the second. I couldn't think about it, I didn't want to, although I knew I would later. I am a firefighter, it's Saturday night and it's only going to get busier.'

9.30 pm Dublin Docklands
The street was quiet when the cigarette started to smoulder in a long-derelict house. By the time the embers had burnt through the film of dirt to the linoleum beneath, night was gathering outside and there was nobody to notice the small wisps of black smoke which began to seep through the holes in the bricked up windows. Feeding on human and animal detritus in a warm environment, the flames remained small but began to spread, lapping hungrily at the plastic coating of the dead electrical cables by the kitchen door.

9.45 pm Southern suburbs
It wasn't until half-time when a crowd had gathered outside the pub for a cigarette that someone noticed the broken beer bottle encrusted with blood and hair. It was five minutes later when a young man, bleeding profusely from a head wound, was spotted lying beside the rear tyre of a Ford Fiesta in the car park. One customer called the Gardaí. The other dialled 999 for the emergency ambulance.

9.47 pm Docklands
It was the frayed dry remnants of the carpet that gave the flames their grip on the staircase on the abandoned home. Beneath, the dry rotting wood soaked up the flame eagerly. The entire lower floor was now a mixture of low-lying flame and a pall of dense black smoke. On the street front, it would still have been hard for the few passers-by to notice the small but steady trails of black smoke escaping from the

bricked-up door and windows. However, to the rear, overlooking a nettle- and rat-infested garden, the bricks had been pulled loose and the smoke travelled freely. Like fireflies, the first sparks fluttered into the dusk sky and settled on the tinder-dry nettle leaves. Upstairs, a man leant against a closed door, a dirty hypodermic syringe jutting from his leg, fast asleep.

9.49 pm Newland's Cross

It was probably the fortunate fact that nobody had been hurt that allowed the jokes on the fire appliance as it waited at the lights at Newland's Cross on the Naas dual carriageway. Although it could have been tragic, the fact that a grown adult thought it was a good idea to jump around on a bouncy castle with a cigarette while intoxicated, did have its humorous side. It was stupid sure, but as most firefighters will tell you, the job makes you realise how much of a potential disaster the general public are.

The lights at Newland's Cross were always bad but the crew were nearly back at Tallaght station. Hopefully, they could grab a few minutes for a bite and a chat. All the way to Lucan just because of someone's stupidity. Just as the lights changed to green, the rising sound of an ambulance siren could be heard. The driver of the fire appliance pulled aside and the ambulance sped through the junction as the lights turned orange. Although the callout to Lucan hadn't required an ambulance, someone needed it now. With only twelve ambulances in Dublin Fire Brigade providing emergency 999 service for the entire city, quiet nights are very rare indeed.

9.50 pm Belgard Road

Three cans of Bud, no problem. Sure wasn't he just around the corner, even if he only had a provisional licence? The Opel Corsa took a turn out of Kingswood estate and was approaching Belgard Road when the driver's phone rang. Reaching towards the passenger seat, he didn't feel his hand turning the wheel to the right just as the Renault Clio was approaching.

Although she was in a 30-km zone, the Renault driver pushed the accelerator a little just so she could get through those orange filter lights to turn right. She was just pushing 50 km when the right bumper of the Corsa came out of nowhere.

9.51 pm

It was the car in front swerving wildly to the right which gave the first indication of trouble to the ambulance crew behind. In the growing darkness, a lone indicator light was flashing at a very strange angle. A figure was staggering uncertainly on the hard shoulder as the ambulance pulled in, crunching over shards of glass.

9.55 pm Docklands

As it turned out, the driver of a large eighteen-wheel lorry on its way to the North Wall to the ferry was the first to stumble across the fire. Taking a wrong turn on the way through the city, he had ended up in this maze of mostly deserted streets. It was the black smoke in his halogen headlamps that he noticed first. Opening the cab door, the smell of burning plastic nearly made him retch. Walking towards the building he could hear the crackle of flames and, above the house, the sparks ebbing and flowing in the night air.

9.53 pm Tallaght

If it wasn't for the airbag, the woman in the Renault would have died. Instead, she had a dislocated shoulder and neck injuries. The man in the Corsa had sustained head and facial injuries. The ambulance, en route to an incident at a nearby pub, now had two casualties bound for Tallaght hospital.

9.54 pm Southern suburbs

A crowd had gathered at this stage around the assault victim in the pub car park. One laid a towel beneath his head while friends cursed the delay in the ambulance arriving.

9.55 pm

The ambulance from Rathfarnham station had just returned from a minor road traffic accident (RTA) on one of the many speed traps on the new M50 when a call from HQ in Tara Street directed them to a pub in the Tallaght area, a 22-year-old man with head injuries sustained in an assault with a broken glass bottle.

10 pm Docklands

The rear of the derelict building was entirely in flames now, the staircase completely engulfed. Fire had also begun to take hold amidst the tinder-dry detritus of the back yard, neck deep in grass and nettles. The dispatch of two units from Tara Street station had taken place just over two minutes previously, and both appliances were weaving their way through the maze of one-way streets and cul de sacs which fill the docklands area. Another motorist had stopped at this stage and was conversing with the truck driver. Both men were wondering where the hell the fire brigade was. The truck driver insisted he'd called them fifteen minutes before. At this stage, a small crowd from a nearby pub had begun to gather at the end of the street, most clutching cigarettes and pints of beer. Upstairs, the fire lapped hungrily at the closed door of one of the rooms. Inside, the man's still prostrate body was now wracked by continuing convulsive coughs as smoke began to fill the room.

10.02 pm Southern suburbs
The car park to the pub was packed now. On the road outside, cars had slowed down to see what was causing the commotion, causing traffic delays a kilometre back down the road to where the ambulance was trying the swerve through traffic at a congested roundabout. At the pub, two gardaí who had arrived on foot were trying to clear a space around the victim when the ambulance arrived. Both paramedics pushed their way through the crowd, asking the gardaí to keep the people back. Someone had placed a jacket beneath the victim's head, raising it at an angle and causing potential breathing problems. There was some blood in the victim's mouth. A gauze bandage was applied to his head wound while a spinal board was slid under him, securing his back and his head. The ambulance left the scene at 10.06 pm.

10.05 pm Tallaght Hospital A&E
At Tallaght hospital accident and emergency department, the heavy price of a Saturday night's excesses was becoming evident. The waiting room was already full of patients when the DFB ambulance from Tallaght arrived. The woman, who had sustained shoulder and neck injuries, had been immobilised on a spinal board as a precautionary measure while the driver of the Opel Corsa was able to walk himself into the hospital clutching a bandage to his face while an EMT applied another to a wound on his scalp. Both would require stitches and one would likely leave a nasty scar on the man's face. After

both patients were inspected by hospital staff, one of the ambulance crew went outside for a quick cigarette. A moment of calm before the next call. Inside, the other EMT gulped down a quick glass of water as he sent a quick text message to his girlfriend. This was his first Saturday night working the ambulance since he joined the brigade. He walked over to a hospital bin to dispose of his blood-stained latex gloves as hospital security personnel tried to restrain two teenagers obviously high on a dangerous mix of drink and drugs. A third lay on the floor, retching. As the firefighter left A&E he noticed that the smell of alcohol was everywhere. He knew it wouldn't be his last visit there that night.

10.04 pm Docklands
The crowd parted slowly as the D/O car heralded the arrival of the two fire appliances, making their way slowly up the narrow street. As the firefighters leapt from their vehicles, prepping hose lines and securing the scene with help of arriving gardaí, random bursts of flame could be seen erupting through gaps in the roof slates. There were very few access points to the building and firefighters used crowbars to try and prise away some of the planks barring the windows so they could deploy hose lines and start fighting the fire. Two firefighters walked around the corner where they could get a limited view of the rear of the house. Hoisting himself up on the wall, one saw the entire back garden in flames and also noticed the unbarred window to the rear. Hoses were soon snaking around the corner as the crews tried to damp down the blaze in the garden before entering the building. The D/O had initially thought that it might not be necessary to send men inside the crumbling, dilapidated structure but he had just been informed by gardaí, corroborated by some members of the public, that the house was used occasionally as a squat for drug addicts. Persons reported. It would need to be searched and time was of the essence.

10.12 pm Tallaght Hospital
On arrival at the hospital, the unconscious assault victim was treated as an emergency. Profuse bleeding from the mouth and head indicated severe injuries consistent with a savage beating. He was prepped for emergency surgery. The two-way radio in the ambulance was already patching through a host of calls for the ambulance crew.

Another stabbing at a nearby chipper, a man had fallen over drunk and sustained head injuries at a bus stop and a cyclist had just been involved in a collision with a bus.

10.15 pm Docklands
Gaining access to the house wasn't easy. Two hose lines had dampened down the flames in the back garden enough to send two personnel over the wall to gain access from the rear but, as with many things in the job, it was easier said than done. At least half a dozen rats had been seen scurrying away from the house since fire crews arrived, which meant there were definitely a great deal more. Hardly a major problem, but an unwelcome distraction, particularly if one of the firefighters was to step on a nest of them while navigating the overgrown back garden. At the front on the building, two more hose lines were hitting the fire through holes created in the ground- floor windows. In order to gain access to the first floor, where the flames were still raging, they would either need to use RTA cutting equipment on the steel which covered the upstairs windows or else attack the fire from the rear, which was the chosen course of action.

The two firefighters clambered through the hole in the rear wall and began suppressing the flames with controlled bursts of water. The flames had engulfed the staircase almost completely, and when one of the firefighters put his foot on the second step, it almost gave away beneath him. Searching the darkness around the base of the stairs by torchlight, the beam reflected off narrow slivers of melted plastic and the shrivelled burnt remains of clothing. The firefighters had been in enough drug dens and squats to recognise the tell-tale signs. Having extinguished the flames as far as the landing on the staircase, one of the firefighters used what was left of the sturdy wooden banister to clamber as far as the landing. Upstairs he could see the closed door, now completely ablaze. Motioning to his colleague for assistance he gestured upstairs and the hose kicked again clearing flame from the upper steps. Outside, the s/o ordered the two personnel who were conducting the damping down operation in the back garden to go inside as backup to their colleagues upstairs. The remaining firefighters began to prep ladders to deploy at the upstairs windows, where they could use power saws to cut through the sheet metal which enclosed them.

10.20 pm Northern suburbs

He'd been spat at before while on callout. However, as the paramedic pulled his hand away from his mouth he thought he had cut himself. It was only then that he realised that the patient who he and his partner had just secured in a stretcher had spat at him. Wiping the smear of blood away from his eye, he tried to retain focus on his job, trying to keep to the back of his mind the fact that he had just been spat at by a known intravenous drug user with possible HIV. The lifts in the flats were out again so they carried the patient down four flights of stairs. This was the fourth time they had been called to the flats that week.

10. 18 pm Docklands

The flames had spread half-way across the upper floor of the house but as of yet had not touched the second upstairs bedroom or bathroom. The water had done its job so far, reducing most of the interior to a blackened, scorched, drenched shell. However, the door was still closed and it was possible that the fire had got inside. The firefighters knew that if there was someone in the building they would likely be in this room, one of only two they hadn't searched to date. However, the closed door represented a significant danger. The fire may have entered the room and while it devoured everything within, it may also have consumed all the air within the room, reducing the fire to an unburnt gas, invisible and waiting for the introduction of air to allow it to blossom again. Realising the danger, one of the firefighters passed the message downstairs to the backup team who in turn alerted the crews with the cutting gear at the front of the building not to attempt entry to the sealed room. The threat of backdraught is one of any fireground's most terrifying phenomena.

Using his gloved hand, the firefighter gripped the brass doorknob, feeling the intense heat through his gloved hand and noticing how the scorched timbers of the door pulsed in and out, similar to a breathing motion. The perfect ingredients for a backdraught.

10.33 pm Tallaght Hospital

One of the A&E doctors gave the firefighter who had suffered the blood spray a quick check-up and took a sample of blood from him. As much as three months of waiting now for the results, relationships and life on hold. The firefighter sat in the A&E department for another two minutes, trying to gather his thoughts and control a rising sense of fear and panic. Then the curtain was pulled back and his colleague was there waiting for him. Another call . . .

10.40 pm Docklands

The scorched ceiling in the kitchen signalled how the fire had gained access to the upstairs room. Now, in that space, it waited. Having consumed all the air within the room the fire itself had been reduced to a mist-like cloud of gas, waiting for the availability of more air so it could burn again. The man slumped in the corner had met his death in the depths of a drug-induced sleep, choked by smoke and seared by the rising temperatures. Outside the room, on the landing, the flames had been extinguished. One firefighter crouched before the door, three colleagues behind him on the stairway, hoses at the ready.

The man with his hand at the door remembered what one of the senior men on the watch had said to him, 'when you open a door, make sure you keep your head down.' It was something that stuck in his head and strangely enough he was still thinking about this as he opened the door. There was a moment's silence before a high-pitched whistle became a roar and the flames shot over their heads, spreading once again over the walls and ceiling of the room. The firefighters advanced, hitting the fire with bursts of water, staying close to the walls where the floor would be strongest. The leading firefighter was conscious of controlling his breathing in the BA, trying to focus on the job, knock down the fire, get downstairs and get back to the station safely. No shocks, no surprises, this callout had been interesting enough already.

When his left hand happened upon something solid he thought it was a piece of fallen wood; shining his torch on where his hand lay he saw what remained of pair of jeans, melted into a badly burned leg. Moving the light upwards he saw the rest of the body, face down. He dropped the torch and jumped backwards. Other torchlights arced towards him, he felt a hand on his shoulder. Someone cracked a joke. Someone had to. He had gotten a fright and he knew that the other guys weren't about to let him forget it in a hurry. One of the other firefighters dragged the body by the legs as far as the doorway. With the flames extinguished, the house was now deathly quiet and dark as the firefighters made their way down the stairs with the grisly remains.

10.50 pm Southern suburbs

The cyclist's head was at a sickening angle as he lay against the kerb. The arriving ambulance crew knew that deploying the spinal board and cervical collar wouldn't make much of a difference at this stage. The headlights of the bus lit the scene in a garishly brilliant light as the crowd pressed closer. The bus driver was talking to gardaí, who were trying to calm him down. The cyclist had come out of nowhere and wasn't wearing a helmet, never stood a chance. However, statements still had to be taken. For the paramedics this was an open-and-shut case. They took the body to the ambulance quickly and silently. Right now, they were undertakers.

11 pm Finglas Station

It was kebab and chips, again. As the new guy on the D Watch in Finglas, he decided it was best to play it safe. Two hours chopping onions and peeling potatoes and cutting chicken. He had been on a callout earlier where he had taken a fair bit of stick about the last meal he'd cooked. Which was also kebab and chips, but he was determined to get it right this time. Not that he'd get a bit of praise for it. Respect maybe, but praise? Forget about it.

He heard the bay door at the station open. The D/O walked into the kitchen and took a quick, quiet look at the grub. He shook his head, grabbed the *Evening Herald* and walked into the mess room. He was joined, in much noisier fashion, by the other lads on the watch. It was Saturday night and they were unlikely to get much time to themselves, so the jokes and abuse flew thick and fast. When the kebabs arrived, followed by trays of glistening crunchy chips, there was a consensus of mock disgust from most of the guys on the watch. The rookie had to grin and bear it. It'd be his turn to dish it out some day.

Of course, like every watch, there was always a loner. He came in last, thanked the rookie for the food and ate quietly. The television in the corner was on, but nobody paid much attention. The conversation changed direction multiple times as wrongs were put to right and any egos put firmly in their place. The kettle had just boiled for the customary dessert of tea and biscuits when the ambulance bell sounded again. The guys on the ambulance left, grabbing a fistful of biscuits on the way, fond farewells ringing in their ears.

11.10 pm Clondalkin

The stabbing had turned out to be a false alarm, but if the lads from Tallaght thought that their night was calming down, they were sorely mistaken. The next call was to a housing estate in Clondalkin. When the woman opened the door, she was crying inconsolably. As the paramedics entered the house they could hear children crying from the kitchen. Walking into the living room they saw a man, presumably the father, lying by the fireplace. Watching television, he had stood up clutched his chest and collapsed. He had barely a pulse and very shallow breathing. The defibrillator was attached to his chest. The family gathered around at this stage, crying and sobbing as their father bucked with each shock passing through the electrodes. His pulse quickened slightly and the crew knew that they had to move quickly. Within three minutes he was on a stretcher and on the way to the hospital. His life hung by a thread right now, but they had done all that paramedics were expected or trained to do. To turn almost certain death into a fighting chance to live.

Midnight

The dark waters of the Liffey shimmered slightly beneath the lights and the *avant garde* glass architecture of the International Financial Services Centre on Dublin's docklands. A testament to economic and social progress, tonight it served as the backdrop to a human's ultimate demise. The man clutched the railing of the bridge and stared down intently.

Earlier in the evening he had driven to the house of his former mother-in-law in an almost trance-like state. Staring through the windows he had caught glimpses of his estranged wife and their two children. He continued to drink, at first to pluck up the courage to try and talk to her and then, when half the bottle was gone, he drank just to forget. To forget everything. He wanted everything finished and that was that. He had started his BMW and driven away slowly, struggling to focus on the road as the alcohol cloyed his senses. He had parked the car just beside the memorial to the Irish Famine next to the Financial Services Centre and now stood next to the railings by the river's edge. The traffic was dying now and there were very few people walking on the quays. He had this moment to himself. Lifting his weight over the railings he teetered at the water's edge, his balance compromised by the amount of alcohol in his system. He clung to the railing and waited. Then, closing his eyes, let go.

The woman on the bus crossing the bridge towards the south side of the city swore later that she saw the man staring at her as he let go and dropped into the dark water. When firefighters arrived, she pointed emphatically to the place where she had seen his body disappear. Less then ten minutes had elapsed. The pitch-black waters of the Liffey were particularly uninviting in late autumn, one of the D Watch lads thought as he finished putting on his SRT gear. Torchlight played across the surface of the water, where nothing moved. A few people had gathered, people with nowhere else to go, around the fire truck and watched with nonchalant curiosity as the firefighter prepared to enter the water.

Swiftwater Rescue Technique (SRT) is one of an increasing number of specialist activities in which DFB personnel are trained. The Liffey is one of the major killers in the city every year, so firefighters trained in this discipline have become particularly familiar with its vagaries and perils. To say the riverbed was unpredictable would be an understatement. Moving slowly, the firefighter edged a traffic cone aside and almost tripped in the frame of a bicycle, which drew the expected round of applause from his colleagues at the dock wall, always sympathetic. His target was what looked like a piece of wood in the centre of the river; he would search that far and no further. It wasn't far, and he reached his hand out to grab it when his shin struck something which moved sluggishly. Signalling to his colleagues, he felt beneath the surface of the water with this hand and, head first, pulled the body clear. Signalling to the crew, they began to help him back to the quay wall.

1.15 am
Dublin City Council staff-support services weren't readily available at this hour of a Sunday morning, so the firefighter who had been spat at while trying to prevent a drug addict bleeding to death sat in the kitchen of Tallaght station and spoke to his colleagues. Amidst the black humour, which acts as a kind of curative for the situation, helping to cleanse the memory of a bad callout, there were words of genuine reassurance and support. Guys had been there before, been through the same thing and they had been fine. Put it to the back of your mind and try not to think about it. Even the guys who were

giving him this advice knew that you couldn't *not* think about the possibilities of such a situation, but they also knew that like anything serious in this job, the sooner you got it out in the open and talked about it the less threatening it would seem and you would find it easier to get on with the job. The official structures within the organisation were there during office hours but it was here, in the station, where the problems were most frequently discussed.

3.30 am Tara Street, HQ

It seemed the city was drawing a breath, finally. The firefighters in HQ hadn't seen the ambulance crew all night, but that was common enough on a Monday, let alone a Saturday. It had been a bad night for 'good' fires, as more severe blazes are known. Apparently, there had been quite a bad one out near Cabra, two little girls caught in a house fire. One of the lads from Phibsborough had gotten them out. One was dead and the latest news was that the other would be lucky to see the morning, but she was still hanging on. The fire in the docklands had been a dirty and nasty experience, and one of the guys was paranoid that some of the scratches he'd sustained might have been caused by rats. Unlikely, but he hadn't been able to get it out of his mind. There were other rumours around the mess table, stories pieced together from the night, picked up from other lads while on callouts or over the radio. The group at the table had thinned considerably. There was a very minor chance that they might get a few hours of uninterrupted sleep at this hour. Very minor and they didn't mention it for fear of jinxing it.

4.45 am

The hospital staff and DFB crew were pretty much on a first-name basis at this stage whenever the ambulance from Tallaght station arrived yet again at A&E. Muggings, stabbings, assaults, domestic violence, drink overdose, drug overdose, overdose of both, all in the lexicon of a busy night. Occasionally the paramedics would see a guy who they had taken in hours ago still lying on a trolley; nothing they could do about it: one end of the service was as overstretched as the other.

5 am Phibsborough Station

To be honest, the firefighter was happy when the bells went again. He had been in bed but he hadn't really been sleeping. That's far worse, of course, because those are the moments when you start thinking. He tried not to think of the night's first callout but it was hard not to. It wasn't so much that he was thinking about it, more the fact that he was trying so hard *not* to think about it. He had talked about it when he had got back; everybody had agreed that it had been a pretty vicious fire. He hadn't seen the lads from the ambulance since, so he didn't know exactly how the first girl was. Surprisingly she had been alive when they got her to the hospital.

When the bells went, his mind cleared and he had something else to focus on. The appliance was quieter then normal. The D/O was talking about a bad fire down the docks. Some junkie found dead in a burning building, apparently quite a fire. The lads from HQ were at it. They were on their way to this fire too.

5.15 am Dorset Street

I suppose if you look at it from a civilian's point of view, every fire close to your house is a major one. Maybe that's why there were over fifteen calls about a fire in a skip off Dorset Street. No fire is to be taken for granted but it definitely didn't need units from both Phibsborough and Tara Street. Nevertheless, there they were, on auto-pilot again. Bin charges can put quite a hole in the average working-class pocket in these areas so under cover of darkness, it had been decided by somebody to rid themselves of an old sofa, several rolls of carpet and some lino. Where the spark to light the fire had come from was anyone's guess.

The guys from Phibsborough were there first and nearly had the blaze out when HQ's truck arrived. This was the sort of fire you could knock down aggressively and they moved quickly, a far stretch from their first callout of the evening. The firefighter in question was feeling better now; his mind had moved on for the moment. Just as they were packing up, he saw one of the lads from his recruit class climbing into HQ's appliance as they got ready to return to station. He hoped that his former classmate had a better night then him.

The guy in HQ's truck knew most of the heads from Phibsborough but couldn't quite place this guy who was waving to him like he knew him. Must know me from somewhere, he thought.

'How's the night going?' the guy said, approaching him.

He shrugged his shoulders, 'the usual, what about you?'

'Ah, you know yourself, Saturday night.'

Something in this guy's face made it clear that he wanted to talk more, so the guy from HQ asked, 'Any bad ones?'

'Yeah, one was rough enough,' came the reply. Smoke, fire and still, small bodies crossing his mind. 'You guys?'

'Bit messy alright,' came the reply from HQ. 'Down the docks, some house with a junkie in it. Messy enough, you know yourself. What about the bad one earlier?'

The engines started as the firefighter from Phibsborough nodded. 'You know yourself, children, bad alright.'

The D/O had started his car and was moving away. The guy from HQ rolled up the window and waved as the truck began to move off. The firefighter sitting beside him, elbowed him. 'What was that all about, he looked a bit freaked. Do you know him?'

'Can't place him but I know him from somewhere. Rough night, I think.'

'Still, didn't look as bad as you, though.'

'Thanks!'

'No bother.'

8.45 am Phibsborough Station

The lads from A Watch were arriving in at DFB stations throughout the city. Saturday night blending into Sunday morning, chalk and cheese in terms of workload. In Phibsborough, breakfast was finishing up. The news had carried details of the night's work. Nobody made a point of it but did keep an ear open for news from the night's bad domestic fire. The girl was still in intensive care. Still critical, still alive though. Of course that's the last they would hear unless she died. The A Watch guys who filtered through the locker rooms knew what Saturday nights could be like.

One firefighter had hardly been able to sleep a wink, the children on the stairway and in the bedroom still on his mind; he hoped he'd be alright to drive. One or two guys offered to drive him from Phibsborough station. He said he'd be OK, needed some space, but he'd be fine. One of his friends said he'd call round later, have a beer and watch the game.

When he arrived home, the kids were up, milling around in front of the TV. His wife saw him come in. 'How was the night?'

'The usual madness, Saturday, you know.'

'Yeah, you sure you're OK?'

He nodded. 'I'll be fine.'

She knew by his face that he may well be fine later, but he wasn't right now. 'ok,' she said. 'Get some rest.'

He went upstairs and showered again, the smell of smoke in his nostrils never seeming to clear completely. Putting on his dressing gown he checked the smoke alarm in his bedroom, then the one on the landing before putting a new battery in the smoke alarm in the children's room. He could hear his wife downstairs, asking the children to keep quiet. He closed the door and climbed into bed, burying his head under the pillow and trying to drown out the voices of children. He needed to sleep, the sleep of the dead, but when he closed his eyes he could still see the face of a child, blackened and shrouded in smoke.

Part One

The Job

Being a Firefighter

So what is a firefighter and how does the fire service work? Today's service demands a wide range of talents and aptitudes from those who pass the test and are deemed able to do the job. Firefighters throughout Ireland, North and South, receive very similar basic training using the same standard equipment and fire appliances and generally the same techniques. There are over three thousand firefighters in the Irish republic, operating from 220 stations nationwide, run by thirty-seven fire authorities. The Fire Services Act of 1981 made provision for a Fire Services Council to be appointed by the Minister for the Environment and Local Government, as it was known then. This Fire Services Council has been active since 1983. One of the main functions of the Fire Services Council is to train fire service personnel. This training is then carried out at local authority level since there is, as yet, no national training centre for firefighters. The Fire Services Council also advises the Department of the Environment on the educational and training needs of firefighters.

Fire brigades around the country, who once solely battled fires, now respond to a wide range of incidents such as road traffic accidents (RTAs), chemical spills and fires, water rescue, general rescue, flooding and a host of other incidents that fall within their remit. In Dublin, firefighters are also trained paramedics (until recently known as emergency medical technicians). Elsewhere in the country, the benefits of pre-hospital training are also being realised with many fire crews being trained to render advanced first aid until the arrival of an ambulance.

Of course, the primary reason why fire brigades exist is to serve as the front line against one of man's oldest enemies. That is why someone chooses this career, finding out in the course of their training and their work that there is often so much more to it than that. The priorities of firefighters are many, the main three being to save life, to save property and to render humanitarian assistance. The classic image of the firefighter, perpetuated through the mass media, plucking a child from the jaws of flame is in many ways true. A firefighter will do this: if there are persons reported inside a burning building then every option will be utilised to save human life. However, fire services in this country also have an enviable safety record and they will not do something reckless or foolhardy. Safety is paramount. A fire officer will not order his/her crew into a situation where they are likely to have more casualties coming out then they had going in. The training which they undergo is designed to make them efficient and effective when doing the job and also ensure that they go home safely at the end of it. Risk is calculated, not guessed.

There are two main divisions of firefighters in this country, the full-time brigades and the retained. Full-time firefighters do the job 24/7 and there are five brigades in the Republic: Dublin City, Cork City, Limerick City, Waterford City and Galway City. All other counties have a retained service. A retained firefighter is paid a quarterly retaining fee by his county fire service, run by the county council, but he/she also has another job in the vast majority of cases and responds to his/her stations for callouts via a pager system. There are fire services in Ireland which are known as retained brigades but actually have some full-time stations. An example would be Louth, which has two full-time stations in Dundalk and Drogheda, the remainder throughout the county operating as retained. There is a similar situation in Sligo, with Sligo town a full-time operation.

Full-time brigades generally use a watch system. A watch is basically the same as a shift. So, in Dublin for example, there are four watches: A, B, C, D. Dublin has 11 full-time stations, plus four retained, and the firefighters on duty in all of them at the same time are all part of the same watch. The watches are planned on a calendar system so firefighters work quite an equal blend of nights and days, weekdays and weekends. The stations can be either 'one pump' or 'two pump' or, in the case of HQ, even more. The number of pumps is basically the number of appliances at that station. Finglas is a one-pump; Dolphin's Barn is a two-pump. All stations will have an ambulance in addition, except Dun Laoghaire, and most will have what are referred to as 'Specials'. A 'Special' is a specialist vehicle designed for dealing with advanced or complicated situations. The 31-metre turntable ladder at HQ or the foam tender for dockland fires at North Strand station would come into this category. All stations have a numerical designation; in Dublin it is 1 to 9, with HQ in Tara Street referred to as 1-0, not ten. In addition, all vehicles have a number. For instance, Dolphin's Barn is referred to as number two. Its fire appliances are known as 2-1 and 2-2. All ambulances are referred to as number four, so in the case of that station 2-4. When being dispatched, the call goes through as 'Delta 2-4'. The 'Delta' meaning Dublin. Elsewhere in the country, similar callsigns exist, for instance 'CK' in Cork and 'GY' in Galway.

Fire brigades in Ireland follow broad international standards in terms of how the brigade is commanded. The chain of command is quasi-military in structure with different uniforms and badges for different levels of firefighter and officer. The model used in the capital is by and large used around the country with some county-by-county modifications, where some ranks may not exist or they may be known by some other name.

The Chief Fire Officer (CFO) is in overall charge of the brigade and this rank is used countrywide. The CFO operates on a managerial and strategic basis for the brigade at county/city council level and they are rarely seen on the fireground. Below him/her is the Assistant Chief Fire Officer (ACFO). There are generally more then one of these officers in any brigade — four in Dublin — their duties are predominantly administrative but unlike the CFO he/she will have worked as a firefighter at some stage. He/she will also be present at the scene of a major fire or emergency. Below them again is the Third Officer who acts as a subordinate to the ACFO.

District Officers (D/O) are in charge of several stations at a time on one watch. The D/O is the highest operational rank within Dublin Fire Brigade and is generally occupied by those with many years' experience within the brigade, working their way up through the ranks. The most senior D/O on any particular watch is the on-duty Mobilisation Officer (known as Mobi). The Mobi's job is to oversee overall deployment of the brigade's resources within a shift.

The next rank down is Station Officer (S/O). The S/O is in charge of one station during one watch and is a vastly experienced firefighter who will travel with his crew on any callout. A full-time station in Dublin will have four S/Os, one for each watch. Second in command in a station to the S/O is the Sub-Officer (SUB/OFF). In a two-pump station the SUB/OFF will be in command of the second appliance, the S/O travelling in the first. A SUB/OFF will also be in command of a single-pump retained station and is qualified to command a specialist unit, such as chemical decontamination. Below him is the firefighter, who carries out operational duties at their superior's command.

There are some differences to the command structure, for example in Cork there are Leading Firefighters who operate at the head of a group of firefighters, and some counties have Second Officers instead of Third Officers, but in the main, the above structure applies to Dublin and around the country.

When a firefighter reports for duty, one of the first things he/she will do is 'parade'. This is where the crews muster for duty and are inspected by the s/o. Each firefighter has a badge number, in ascending order according to experience and his/her badge numbers will be called out assigning him/her to a particular fire appliance and a particular position on that appliance. For example, on a standard fire appliance, number 1 position is the driver with numbers 2 and 4 as the first breathing apparatus (B/A) crew. There will also be two crew assigned to the ambulance. Next, and most importantly, is the equipment check. Here a crew will perform a series of tests on the vehicles and their functions and will also do a stock check on the vehicle's equipment and its condition. One of the most important tests is on the B/A air cylinders themselves, where any problem could be potentially fatal. The array of equipment stocked on the standard fire appliance in Ireland is quite breathtaking and gives yet another insight into the amount of work which firefighters are now able to do. Of course, there are modifications from service to service but the following would be standard on the majority of fire appliances:

Firefighting Equipment
— 13.5-metre, 3-section aluminium main ladder
— 10.5-metre aluminium 2-section main ladder
— 5.1-metre folding aluminium roof ladder
— 4.5-metre 2-section short extension ladder
— Breathing apparatus tally board
— 4 suction spanners
— Foam induction tube
— 4 lengths of 2.44-metre hard suction
— 4 shovels
— 1 wrecking steel bar about 1700 mm long
— 1 ceiling hook
— 2 flaked lengths of 45-mm hose with a spray branch
— 1 foam-making branch (450 litres minute)
— 1 foam solution inductor to pick up foam from a container
— 19-litre fire extinguisher
— 19-litre dry powder fire extinguishers
— 2 carbon dioxide fire extinguishers

— 1 portable water pump to obtain water from rivers etc. inaccessible to the main fire engine
— 1 dividing breeching for splitting one line of large hose into two
— 1 collector head for the portable pump to connect main hose supplying water to it
— Breathing apparatus guideline tags and tabard for officer
— 4 x 45 mm-diam. hose 25 metres long
— 1 bucket and shovel for chimney fires
— 1 x 6-metre length of 100 mm hose for ships and large mains
— 2 x 55-metre first-aid hose reels attached to engine
— 1 x stirrup pump for chimney fires
— 2 x sets of chimney rods
— Salvage sheets for protecting people's carpets!
— Suction basket and strainer to sieve water going into the pump from a river etc.
— 1 drag for pulling out rubbish from fires
— 1 large yard sweeping brush
— 2 gorse-beaters (long poles with flexible rubber mat on one end)
— 10 lengths of 70 mm-diam. 25-metre large hose
— 3 water branches, all with spray-diffuser pattern and .3 mm droplet-making capability
— 2 hose becketts/slings to carry the weight of vertically suspended hose

Protective/Medical Equipment
— 6 chemical-incident suits or 2 disposable gas-tight suits
— 2 dry suits
— buoyancy aids and associated equipment for Swiftwater rescue operations
— 2 sets of waders
— 6 lifejackets
— Hazardous chemicals reference books
— Gas detection unit
— Hand radios
— Intrinsically safe torches
— First Aid items
— Physiocontrol 500 defibrillator
— 1 x Set of 6 Stiff-Neck immobilisation collars
— 1 x small oxygen cylinder

— Dressings
— Water gel for burns
— Sphygmomanometer (blood pressure cuff and gauge)
— Stethoscope
— Selection of oxygen masks and airways.
— Blankets
— Adult and paediatric bag valve mask

Rescue Equipment
— Selection of ropes, including a lowering line with two loops at one end
— 1 x 50-metre dropline which crews carry into a high building and drop out the window to hoist up hose etc.
— 2 x 20-metre floating throwlines and bags for water rescue
— Set of insulated poles and earthing cable equipment for use on the Luas light rail overhead cables in Dublin

RTA Equipment
— Set of wooden cribbing blocks for stabilising cars in road traffic accidents
— 2 large, high reflective road traffic accident signs
— 4 rubber hose ramps to allow cars to drive over hose
— 1 x large angle-grinder con saw, petrol driven
— 1 x battery-operated reciprocating saw
— Tool box
— 2 sets of electrical gloves
— Glass management kit for vehicle windscreen including Glassmaster saw and soft patient protection
— 2 x 'teardrops': hard plastic protection for patients while in a vehicle being cut up
— A Holmatro pump with 2 tool outlets (720-bar operating pressure)
— 1 x cutters
— 1 x spreaders
— 1 x large ram
— 4 blue flashing beacons for road accidents
— 1 manual back-up hydraulic pump for extrication equipment
— 2 airbag-protection devices to prevent late airbag-deployment on a rescuer/casualty
— Set of collapsible road traffic cones

— Searchlight on 90-metre reel
— 2 x 20-tonne lighting capacity airbags and control unit
— Warning tape
— Spray can for vehicle position marking

Other Miscellaneous
— 1 county-type standpipe
— 1 ball-valve-type standpipe
— 1 standard lug-type standpipe
— 2 keys and bars for opening hydrants and turning on water
— 1 water key
— Various size false spindles for hydrant valves
— Box containing hand axes, landing-wheel valves (water supply in high rise), ship-to-shore hose connection, lift keys, gas keys, and various hose-connections.

After the equipment has been properly checked and maintained there is plenty to keep the firefighter busy. Of course, once on duty, the primary function is to respond to emergency calls. But in between callouts, the station is essentially home and must be treated like one so there is cleaning, maintaining, cooking and other duties to be done. Crews on duty during the day are generally kept busier in this respect and they will likely also carry out drills and training or lectures should new equipment or practices be introduced. But when the bells go, everything stops. Firefighters in full-time brigades are expected to be on the road in less then a minute. Retained services have a window of several minutes, which gives their firefighters time to report to the station from their homes or workplaces.

When responding to a callout, also known as a 'turnout' or a 'shout', the crews will have a certain amount of information as to what they are responding to. This is particularly important if it involves something like a chemical spill or water rescue, which requires separate equipment, protective clothing and other vehicles. It is the same for ambulance crews who will know what sort of medical emergency they are responding to when they leave the station. This information is ascertained in the control room where, using highly advanced equipment, control officers are trained to learn as much as possible about an incident when the call comes in. They are also able to utilise the brigade's resources to send appliances to respond to particular incidents. This is done with the state-of-the-art STORM (Strategic Tasking Operational Resource Management) system, installed in 2004. Dublin is also the headquarters for CAMP (Computer-Aided Mobilisation Project) which deals with calls for most of Leinster. Similar call-taking and call-tracking procedures are implemented elsewhere in the country. The STORM system has greatly increased the ability of controllers to elicit the required information and maximise the efficiency of the response.

In the case of a fire, the s/o or SUB/OFF will retrieve a docket from a printer in the muster or locker room. In the case of a fire it will have information like the following:

House on fire
Persons reported
205 Emmet Road, Inchicore
Dublin 8
Location: junction Emmet Road, Tyrconnell Road
Further details: Three persons trapped in upstairs room at rear
of house. Roadworks in operation on Tyrconnell Road.
Vehicles responding: D21, D24, D91

Other information is communicated via radio to the crew when they are en route to the scene.

On a fireground, the environment can be unforgiving. In addition to the pressure of dealing with a fire, the firefighter will also have to deal with heavy equipment, possible problems in terms of water pressure, the danger of the fire itself, thick clouds of possible toxic smoke and a highly emotive atmosphere for those whom the fire is affecting, particularly if there are people trapped inside. For the firefighter, every callout is given his/her complete attention. If your home is on fire they will work for you to save it. However, back in the control room there is likely a backlog of callouts with the control officer waiting for a certain appliance to become free so they can reassign it to another callout. The firefighters and officers on scene know this and that's why they work with such automatic efficiency and try not to get too emotionally involved, where possible, with what is going on around them.

Training

The discipline inculcated within the firefighter is learned mainly through experience, but the training required of a full-time firefighter is arduous. The selection process is rigid as recruitment classes are well over-subscribed for all full-time brigades. (For the retained service, the selection process and training is very similar but due to the nature of a retained service, some fire services have had trouble recruiting.) In the wake of the Stardust tragedy in 1981, amongst the recommendations of the subsequent investigations and inquiries was the establishment of a national academy for firefighter training. Sadly, twenty-five years later, this has still not happened. However, although most brigades have their own training facility, broadly speaking, there is an excellent working relationship between the various brigades and shared training is common.

Whether full-time or retained, the level of training is of a very high standard. Experience adds another level to the training but there is no cultural bias between the full-time and the part-time brigades. For instance, most of this country's horrific RTAS occur in rural areas. Therefore, in the vast majority of cases, it is retained firefighters who are used to responding to such incidents, acquiring a level of experience in dealing with high-speed RTAS and their consequences that a full-time firefighter in a city centre station may not. In Dublin, the O'Brien Institute (OBI) was purchased in the Eighties as the DFB's dedicated training academy. On average, two to three classes of recruits graduate each year before being assigned to stations around the city.

As mentioned earlier, the core skills taught in services are very similar but recruits for the full-time and retained brigades would obviously have different training patterns. In Dublin, recruits train for a solid sixteen weeks in firefighting and rescue techniques. In general, when the recruits arrive at the OBI, they are divided into different batches of five or six firefighters so they can train together as a team. In this team they learn the basics of what is required to fight fire, once known as firemanship. They do a series of drills to learn these techniques, an almost military form of training which will quickly teach a recruit how to do a series of tasks in an efficient and safe manner. For example, if a crew should be called to a domestic fire in a standard semi-detached house, the recruits learn the discipline required to do the job properly. Just as in a real fire station, they will have a pre-assigned number and seating position at the start of the drill, they will know what exactly they are expected to do once the command has been given. In the case of a house fire, the instructor may command that they 'make down' a 13.5-metre ladder. 'Make down' is a common firefighting term which literally means to deploy a piece of equipment. Knowing their tasks, the firefighters will then go to a position on the ladder and erect it as they have been shown. On a fireground, this will mean that confusion will be either non-existent or kept to a minimum. If there is confusion it will be from external factors, not due to the firefighters' training: there will not be four firefighters rushing to do the same job while something else may be neglected. About half of the training is about learning the basics of firefighting in drills such as this. Other tasks which the recruits will become very familiar with include:

- Making down hoses (small and large) to a fire from the fire appliance.
- Making down hoses from a water hydrant to the fire appliance and then to a fire, or up a ladder and into a building.
- Making down the two main standard ladders, 10.5 metres (35 feet) and 13.5 metres (45 feet) respectively, to structures of various heights and also in confined spaces.
- Making down the suction hose to either a portable pump or to the fire appliance. This procedure allows water to be taken from a river or lake using atmospheric pressure to push the water to the pump. This is a not-uncommon exercise during protracted firefighting operations.
- RTA drills in which work is being done even before the appliance stops at an incident: this includes putting on high-visibility clothing and manoeuvring the vehicle into a 'fend off' position of approximately 45 degrees, to protect the accident scene.

These are just some of the standard drills which may be modified or specialised in by individual brigades. The other parts of the 'job' are instructed through a set series of modules. One of the most important for the firefighter is the Breathing Apparatus, or BA, course. This is a discipline which is totally fundamental to modern firefighting. Properly trained and properly used, a BA set will allow a firefighter to enter a smoke-filled building to rescue those trapped inside and to extinguish the blaze. It is also a strenuous part of the course. For their BA training, DFB recruits undergo a series of incremental training tests, each getting tougher and more strenuous. To be firefighters they will be trained to control claustrophobia, fear, panic and stress. All these factors can seriously affect a firefighter in BA on a fireground.

The air contained inside a standard firefighter's BA set is just normal air which has been compressed to fit inside a cylinder. The cylinder which firefighters wear on their backs has a 9-litre water capacity, which translates as 1,800 litres of air and the contents of this cylinder will, temperatures notwithstanding, allow a firefighter to work in a toxic atmosphere for approximately forty minutes, if he consumes an average of forty litres per minute. However, and this is what is so important in the training, if a firefighter is doing extremely strenuous work they will consume their air supply a lot quicker. Panic and stress will also exhaust the air supply. The firefighter carries a contents gauge which allows him/her to monitor how much air is remaining and when he/she should leave the building. The facemask of the BA set comprises a polycarbonate high-visibility visor with inner ori-nasal mask which is designed to prevent the visor from misting over. The mask itself is obviously tight fitting and made from neoprene rubber or silicone and held in place by a flexible adjusting harness. Another feature of BA sets which recruits learn is that they are now positive pressure, the air being forced into the facemask. This comes into its own if the facemask is dislodged and the firefighter becomes threatened by incoming toxic gases. The positive pressure of air will help protect them.

BA procedures are strictly enforced in all Irish fire services, firefighters are not allowed go into hostile atmospheres without all the proper equipment and procedures in place. In an atmosphere so inherently dangerous, safety standards must be similarly unforgiving. Thus, aside from his/her air cylinder and facemask one of the most important devices a firefighter can have is the Distress Signal Unit, or DSU. There are various models of DSU in operation but they all have the same function: if activated the device signals that a firefighter is in trouble and urgently needs assistance. This device is activated by withdrawing a magnetic key from it prior to entering a BA environment. The key is attached to a marker known as a tally, the tally being handed to a designated member of personnel who acts as Breathing Apparatus Entry Control Officer (BAECO). The tally is attached to a breathing apparatus incident board. Some of these boards are now computerised, some are manual. On the tally is the firefighter's name, the time he/she went in, the air contents of their cylinder and their station. This information allows the BAECO to determine which firefighters should be coming out and who they are. The DSU will activate if the device does not sense any movement from the wearer for thirty seconds, when it will go into pre-alarm mode. After a further ten seconds, if there is still no movement, it goes into full alarm mode. If it goes to full alarm it cannot be deactivated until the firefighter retreats outside and the magnetic key is reinserted. This device means that an unconscious firefighter can effectively signal him/herself as needing assistance.

Recruits on the BA course will have to conduct a number of BA exercises in confined spaces such as tunnels and small rooms. They may be filled with smoke and visibility will be very poor. In these environments the recruit firefighters learn how to control their own bodies' responses to these hostile environments. When inside toxic environments, recruits are trained to remain in personal contact with the rest of their BA team, be it a minimum of two or a maximum of five. They do this with the 'personnel line', a six-metre extendable rope in two sections, which ensures that firefighters do not get separated while working together.

In addition to the BA training, the recruits are now also taught as standard a course in fire behaviour. This is a recent and much-needed addition to the training syllabus which mixes science into the standard training to create a more knowledgeable and therefore more effective firefighter. The fire behaviour course will teach students what tactics to use against flame in its various forms. In varying situations, fire can have quite a frightening capability to morph into an array of lethal forms. Recruits learn the speed with which flame can move, the fashion in which smoke can explode and also how things can burn without flame. When this module is completed, the firefighter will know how to recognise and deal with flashover, backdraught and other similar phenomena. Fire behaviour is not taught nationwide yet, as there has been no guidance from the Department of the Environment which is ultimately responsible for Irish firefighting.

Recruit firefighters also receive training in pump operation and dealing with RTAS, which often involve complicated, and at times nasty, extrication procedures using powerful tools. This is a fast-growing part of a firefighter's work where skills, equipment and techniques need to keep pace with technological developments. Firefighters are trained how to make safe cuts in different types of vehicles, bearing in mind the new materials used in modern cars where dangers exist for the rescuer as well as the driver. Careless use of an extrication tool could see a firefighter cutting into the airbag canister which can explode. Swift extrication is all about rapidity and patient awareness.

The nature of modern chemicals and hazardous materials has added yet another tier to the training for recruits and the work of firefighters. Trained in dealing with hazardous chemicals, firefighters are capable of dealing with incidents such as spillages and fire involving these toxic substances. The dangers can exist both in the everyday domestic setting and large scale incidents.

Aside from these elements, recruit firefighters are also trained in other elements such as the nature of electricity, building structure theory, physics, marine firefighting and the usage of all the equipment on the fire appliance and the special vehicles. Individual brigades have their own training extras too: due to geographical location some brigades will place particular emphasis on certain disciplines; coastal counties may train in greater detail in marine firefighting for example while midland counties which encounter a higher frequency of RTAS may invest greater training time and resources in those areas. SRT — Swiftwater Rescue Technique — is growing rapidly amongst brigades.

By and large, it is accepted in this country that firefighters and the fire services do a good job. However, it is also recognisable that it has taken major tragedies such as Stardust, Noyeks and Whiddy Island to effectively drive change. In 2002, a major review of the service was published under the auspices of the Department of the Environment, supplied by Farrell Grant Sparks consulting. The results of the report was generally positive news for the firefighters but far more critical of management at county and national level, particularly in relation to emergency management, training and the actual numbers of firefighters per capita, which is lower then anywhere in the UK or Northern Ireland. The report said that on average, overall Fire Service expenditure was 30 per cent less then average spending in EU countries sampled.

Station Life

'You could fill one book with stories from only one watch in a single fire station in this city,' says now-retired District Officer Jim 'Swill' McDonald. 'However, and this is very important, often the stories that firefighters tell are only for other firefighters to hear, they aren't for outsiders' ears.'

D Watch in DFB HQ Tara St is on the night shift, it's Tuesday and so far a quiet night. 'Remarkably quiet,' says one of the senior officers. But it's not a night without tragedy. In West Dublin, a 24-year-old man recently sacked from his job has set himself ablaze in the garden of his former employer. (He doesn't survive.) On a railway bridge near Drumcondra, a woman beset by a panic attack has thrown herself off the bridge. Rail traffic has to be stopped in the area while the woman is rescued by DFB crews. She is rushed to hospital. She'll live. A quiet night so far, but as the clock moves past midnight into Wednesday morning, the crews of D Watch know that there is still a long way to go till dawn. In a city with a population approaching 1.5 million people, tragedy, madness or mayhem are never too far away.

Upstairs, the men in charge of the watch, seasoned Dublin Fire Brigade men with decades of experience, District Officer Paddy Flood, District Officer Jim McDonald (now retired) and Mobilisation Officer Kevin Monaghan, speak of a sixth sense. 'You never know what fire is going to do. The second you think you know what a fire is going to do, you're in trouble. You can try to predict it, but you'll always lose. But, and any experienced firefighter will tell you this, you do become attuned to the various elements present on a fireground, how the fire is behaving and how it's moving. It is like a sixth sense, impossible to explain unless you actually go through it yourself.

'I remember being in a fire with two other experienced lads, up on Grafton Street, must be almost twenty years ago. The shops to the side of Switzer's as it was then (now Brown Thomas) had gone up in flames. We were in a shoe shop, making our way towards the back of the premises. I remember stopping and feeling something, something strange. There was something wrong. I looked at the other lads and they were looking at me in exactly the same way. We pushed the panels on the ceiling up and saw that the flames had burned right through the roof between the two buildings and was now behind us, obviously a highly dangerous situation. I've more stories than I can remember, where I felt that feeling. I remember being in a house fire out in Clontarf, walking through the rooms, hosing down the walls. Then, for some reason, I stopped. A 50-kilo water tank came crashing through the roof above me and crashed straight through the floor to the storey below. If I had have walked just a few more paces, it would have killed me.'

For many years, prior to the advanced understanding of and acceptance of fire behaviour, fire was fought in a more pragmatic manner. In Dublin Fire Brigade for example, until the importance of more structured training became apparent, it was often done on the 'job'. A recruit to the brigade, which was radically smaller until it was increased in size in the mid-Eighties, was inculcated into a society which imparted its knowledge in an almost paternal fashion. The importance of experience, of the 'Senior Man' on a station watch was vital. Today, even from firefighters and officers who entered the brigade following a more structured training regime, this level of experience remains vital. The 'Senior' man reflects the culture of the Watch. He may or may not enjoy a certain unofficial pecking order in relation to duties within the Watch, something which may be bequeathed to him due to the sheer amount of years he has served. It's the same throughout the fire service and especially strong in Dublin, which has nine full-time stations.

Out in Tallaght station, on D Watch, Gary Hayden is one of the more recent recruits to the brigade. 'It's a great place to learn here,' he says, 'there are a lot of experienced guys around. It's not overstated or anything but to have a seniority structure is very important in a station. It's kind of a code and it goes by badge number. There is no official difference in rank or anything, but the fact that another man would be far more experienced then me means that he would have certain informal privileges and would also take the initiative in certain situations. It's nothing to do with arrogance, but it is the way a fire station structure works. It's also been vital in my development as a member of DFB to have senior guys to listen to, you learn so much from them about how to deal with all the different situations we encounter.'

Dave Foran, one of the station officers on D Watch in Tallaght: 'Your badge number is your rank in many ways, you look at a guy's number and you know where he fits in at the station. It is important to have a structure of respect within stations. Terry O'Connor, who has served in Tallaght for almost twenty-eight years, agrees, however,

'While there is a culture of respect, I think there has also been a change of attitude amongst the senior guys in recent years.' There was a time when they had an almost untouchable status, which he believes was wrong. 'Like everyone else in the job, a senior guy still has his work to do. If he starts throwing his weight around, it's unlikely to get him far. It's the same if someone with less experience doesn't respect the senior men within the station. It works both ways.'

The senior man's position within the Watch is driven by respect which he has earned. 'Culturally, this paternal structure of the older members passing on what they have learned to younger members is a reflection of the idea of the station as a house, and the people who work there as a family,' says Paul Carolan, DFB Health and Safety Officer. 'People may think that such a notion is sentimental and misplaced in today's society, but while it has been diluted somewhat, it remains an important part of the identity of not just Dublin Fire Brigade but firefighting as a profession.' There are many reasons for this, one of which is that men, and now women too, may work together in close quarters in frequently dangerous circumstances for decades. There are men within Dublin Fire Brigade who have served on the same watch with other firefighters for over thirty years. 'You have to remember that in some cases, that's a lot longer then many personal relationships might last. There is a bond amongst firefighters, there are also arguments and rows and personal dislikes and rivalries amongst firefighters just like any family because essentially that's what the fire station still resembles, a family house.'

On the safety front, Carolan also believes that this almost familial bond contributes heavily to the enviable safety record in Irish firefighting. 'Without a doubt it does. I remember personally after I graduated that I learned a lot from just watching other people on callouts. It's essential that you do, because the first callouts for a new recruit can be frightening. You could find yourself trapped in thick black smoke, wearing a tight, uncomfortable BA set on your very first callout. The more experienced firefighter will teach you certain tricks and trade secrets to make the job easier. For instance one of my first major fires required us to be on the fireground, fighting the blaze for five straight hours. In the training centre, I had been taught to hold the fire hose in a certain way, but out on the fireground it was very tiring. I remember one of the more senior firefighters coming over to me and showing me how to bend the hose and make it easier to hold and control. It was one of the first lessons which highlighted how important experience is to how we operate.'

It is over sixty years since any firefighters from the capital died in a fire, and the only firefighter to die in operational duty was Timmie Horgan in an ambulance crash in 1994. 'We do have a strong culture of discipline when it comes to safety,' says Paul Carolan. 'There is also a strong element of workplace support which I've said already. There is also a fair degree of that essential element, luck, and we have been exceptionally lucky. However, our safety record is there to be broken every time the bell goes for a callout. We have to remember that. No matter what the callout entails, we need to respect the dangers inherent in the job we do.' The procedures which firefighters in this country adhere to are designed to bring them all back to the station in one piece: they are methodical, deliberate and conscientious in the vast majority of cases. A lapse of discipline is not seen as bravado, it is seen as a reckless gesture which endangers the lives of fellow firefighters.

Either way, even if firefighters do their job perfectly there are other more sinister elements which pose a threat to them in the course of their duties. Violence against fire crews is a problem, not just in Ireland but throughout other countries in Western Europe. It doesn't matter what station in what area, today's firefighter knows he or she could be attacked any time they leave the station.

'You need to watch your back out here,' says Bob Murray of D Watch Finglas, as he and his colleagues often get callouts to Ireland's most infamous, and now condemned, high-rise experiment. Violence against fire and ambulance crews is a real and present danger. Times have changed, the drugs scene in Dublin has changed and improved in some areas, while in others the decline seems terminal. While the pitched drug wars of the mid-Nineties have died down, the area around Finglas station continues to be riddled with crime and the threat of random violence.

'It happens often, far too often,' says Finglas s/o Paddy Hughes. 'Often, when responding to certain areas, we don't even deploy our blue emergency lights. It makes the fire appliance a very easy target.' Quite what makes people target emergency personnel is not debated, it just happens. Finglas is known as 'Fort Apache' in DFB circles.

'Today, there are no limits to what some of these people will do — none at all,' says Finglas FF Colin Murphy. 'Today's criminal in this area, regardless of age, has a disposition of random aggression and violence. Even during the heights of the drug problems in the Nineties there was some semblance of a code amongst these people. The code was that you would not attack or criminalise your own neighbourhood. You had your patch and the next criminal had their patch and so on. Now, the criminals, drug dealers and "junkies" in this area are complete wildcards, capable of anything.'

The fact that you show up in a fire appliance or ambulance, clearly en route to assist somebody, makes precious little difference in the harder and faster areas of Dublin or other major cities. 'I've had to bring people to hospital after they've spat blood in my face. I've also been stabbed with a used syringe' says Colin. The D/O in the area, Jim 'Swill' McDonald (recently retired), says that the level of abuse and the ferocity of attacks directed at emergency personnel is nothing short of shocking. 'They either attack the ambulance, the fire appliance or the personnel themselves. We've also had rocks thrown at the station itself.' Dave Connolly, like Colin, has also been the victim of a blood attack, stabbed with a syringe. 'You have to take them to hospital as they've called you out in the first place. They then get violent and could do something like stab you with a used syringe. Nevertheless, you have to make sure that they're delivered safely to hospital. Then you can worry about yourself. The doctor will then examine you and they will then probably need a blood sample from the attacker to test his or her blood. The thing is, they can only take a sample if given permission from the subject. Meanwhile, in my case, I had to endure several weeks of hell while waiting for my blood test results. It was constantly on my mind. I was nervous with my family, worried about using cups or cutlery that they might be using. It's an incredibly distressing event for anyone. What makes it worse is that it's something caused during the course of your work, helping someone who needed help. That's the reward I got in that case.'

In Dolphin's Barn, on the tough fringes of the south inner city, there are plenty of similar stories. 'Before this station was fortified the way it is now, and I mean fortified, we had huge problems. Basically, we would get called out on a hoax 999 call and they would come in and steal from the station,' says Gerry Sterio. 'We've had occasions here where we've actually come back from a callout and met guys carrying a television set out of the station!' These days, in Sterio's opinion, the key to avoiding the potential risk of violence and attack is common sense. 'A big problem in urban areas such as ours is that people would set bonfires in the middle of open spaces, often surrounded by flats, and when we arrive you get hammered by stones and bottles. So the question then is do we let this fire, which is hurting nobody, burn out so we don't get attacked, or do we do what are supposed to do and extinguish it?'

Gerry Sterio, Peter Charles and George Harrison, who have served in Dolphin's Barn for longer then they care to recall, all remember the extra duties which were part and parcel of firefighters' work in the not-too-distant past. The station is surrounded by several different types of neighbourhoods. There is the South Circular Road and Harold's Cross, both up-and-coming areas along with Crumlin. Then across the grand canal, a stone's throw from the station, is Fatima Mansions.

Keith Carolan, who works on B Watch in Dolphin's Barn says it became part of life at the station to look out for people who may become victims of the area's more sinister elements. 'It was before my time, but it was a much rougher area. They used to call this station "home of the heroes". I can imagine it was said with a fair degree of irony given the sense of humour in the Brigade, even though it was probably true.'

At Tallaght station, dealing with disorder and violence is still part of the job but some believe there is hope on the horizon. 'It still happens, but I think it has decreased out here,' says Gary. 'I mean, it's still a risk, but compared to what some of the more experienced lads have told me it's definitely on the decrease.' Burning cars are a regular occurrence and a regular trap for firecrews. When the fire appliance arrives at the scene, generally a track or a laneway, a vehicle is then moved in behind them and they're attacked. 'At one time we were on average going through a windscreen a week and whatever else would get broken. The vehicles themselves are pock-marked with the amount of stuff that's been thrown at them.'

In Tallaght, there are also personal safety issues for firefighters. In addition to indiscriminate attacks on the brigade and its equipment during callouts, there is also the threat of personal physical violence. 'I have experienced that,' says Terry O'Connor. 'I've had situations where you find yourself in the sort of area where you wouldn't go by choice, when you're in there you're on your own. There may be barriers which would stop the ambulance being able to drive in so you may have to walk to collect a patient. There could be a situation where you could get attacked or your patient could be attacked. You feel responsible for your patient, though, and will likely try to defend him or her if someone tries to inflict further damage upon them.'

Like many firefighters, Dave Connolly at Finglas displays a cynicism wrought with black humour and bitter experience. 'We're here to do a job, which is to rescue people when they need it or to help them when they need help. We exist solely for that purpose. There is no political agenda. You dial 999 and we effectively become your employees. That's why I do this job. I like to help people. However, you can't do this job and not become somewhat cynical about humanity at times. Some of the things people do to themselves and as a result the danger that they pose to other people makes you realise that the general public is an accident waiting to happen.'

And when it does happen, it's firefighters and other emergency personnel who pick up the pieces. Paul Carolan believes that the ongoing attacks against fire and rescue personnel reflects the perception which people have of firefighters. 'I don't believe that firefighters in this city or this country are undervalued, but I do believe that the very fact that we are attacked in the course of trying to help other people shows that we not enjoy the public profile and standing which firefighters in other countries do. Firefighters in the US or Canada do not have the threat of attacks hanging over them every time they answer a callout, it simply does not happen there.'

CISD

Of course, this stress all adds up and can develop into what was dismissed as nothing more then 'part of the job' years ago, but is now recognised as CISD or Critical Incident Stress Disorder. All available psychiatric evidence suggests that responding to critical incidents can trigger emotional reactions in the mind of a firefighter which can be extreme. Left unchecked, these reactions can have a strong and lasting impact on a firefighter's future. Without professional support, emergency workers can be left with permanent mental damage which can negatively effect their personal relationships and can seriously impede their ability to do the job. These critical incidents can be once-off situations which go far beyond what is normal or expected in firefighting. They can result from a particularly challenging or gruesome callout, from exposure to violence, death of a colleague or witnessing a multiple casualty incident. It is a facet, and in many ways a strength, of the job that firefighters remain emotionally detached in the course of their duties. When fighting a fire, or extricating a serious casualty from a nasty RTA, their training takes over and their thoughts and actions are totally employed in getting the job done. Even at this point, however, the process of stress has already begun even though it may not be apparent until the appliance returns to the station, the kettle is on and the crew sit down to talk about it.

In 2001, a conference was held in Ireland on the subject. One of the world leaders in this field, Dr Jeff Mitchell, is a former volunteer firefighter from America. At the conference he spoke of particular traumatic incidents which caused difficulties for firefighters. Any of them could be from any fire station in Ireland. 'There was a man in my unit,' he said, 'A few weeks after our training ended, he was called out to a fire which resulted in the death of a child. A few days later he left the fire service, this surprised me as he always seemed a very strong individual. Later, I learned that his wife had just given birth to their first child and when he was at this tragic fire he kept thinking about his own son, imagining his face in the same situation. I thought it a shame that we could lose such a good man like that.' He went on to say that firefighters can handle thousands of cases, many of them difficult, with no apparent problems and then one particular incident can be the straw that breaks the camel's back. 'Firefighters are at such high risk of post-traumatic stress disorder, occupationally, and yet there didn't seem to be anything in place to help them for many years.'

CISD, like any illness, has a clearly defined set of symptoms which affect the way people think, the way they feel, the way they perform. Firefighters, or anybody, who suffers from CISD often begin to feel isolated from other members of the crew. This can lead to further social problems in the station, on callouts and also at home. A firefighter may begin to feel confused about the job they're doing and doubt its value. There may also be physical symptoms such as stomach upsets, headaches and hand tremors. Firefighters affected have also imagined family members in the place of victims they've seen and as a result often feel overprotective of them. One Irish firefighter said that as a result of what he had seen at fires he had gone to the extent of putting smoke alarms in every room. Other recollections of affected firefighters include the avoidance of places and people associated with the event and an inability to recall details of the trauma. Firefighters can experience estrangement, detachment, sleeplessness, irritability and emotional numbness and distance. Mitchell used a lightswitch analogy to describe the effects: when the light switch comes on and doesn't go off. Treatment for disorders such as this varies: almost every firefighter says that the first port of call, as detailed in their stories, is the station mess room. The experiences of other colleagues, on the same level, can help a firefighter relate to traumatic or upsetting events within the confidential walls of a station. On a formal level, one of the best known techniques is Critical Incident Stress Debriefing, a group-intervention tactic initiated by the s/o which is essentially a structured series of conversations and discussions about a particularly traumatic event and is designed to accelerate the recovery process. The debriefing is designed to work best within hours of the event taking place, before traumatic and damaging memories can become fixed in the firefighter's mind. According to Mitchell, this group debriefing is not therapy. 'It's designed to stabilise the situation, accelerate the recovery process and identify individuals who might need more help.'

Experts believe that one of the keys in helping firefighters avoid long-term damage in the wake of a traumatic incident is by reinforcing, to the firefighter, the fact that what occurred was not his/her fault, however awful it was, he/she did not play an active role in causing the damage. In the wake of stressful incidents it is vital that firefighters don't accept responsibility for another person's tragedy.

While most firefighters still rely on the understanding of their colleagues in the station, the fact that there is now a formal structure in place shows that there is an acceptance of traumatic stress and the dangers it poses to firefighters and, subsequently, the fire service and there is professional psychological assistance available. For the system to work, firefighters need to recognise that they are vulnerable people in a high-risk occupation. They need the maturity to realise and recognise the signs and symptoms of traumatic stress and to seek out support. It also requires the vigilance of colleagues. If someone starts displaying behaviour which is rapidly different from their normal demeanour, then traumatic stress could be involved. In a fire station, traits and personalities are very strong. If someone who is usually gregarious and outgoing becomes very subdued, then there may well be something which has affected their state of mind. The previous attitude of just 'getting on with it' will get people through a difficult period of time but will also likely result in long-term damage to that person's mental health.

Of course, dealing with death is part of the job and a firefighter can encounter scores of cases where life has been lost. Post-traumatic stress may not come into the equation for one particular firefighter but it may for another. Some, like Alan Finn from Rathfarnham station, say that the job gives a snapshot on the state of society and how death is accepted. Recently, Finn and his colleague were called in the ambulance to St Stephen's Green shopping centre in Dublin. In one of the nation's busiest retail centres a man was lying face up on the ground, blood pooling from the back of his head. He had fallen from the top floor of the centre and was killed instantly when striking the hard marble below. Nevertheless, when the crew arrived, many people were walking nonchalantly around the man who lay dead at their feet. 'We understand that people aren't trained to deal with situations like that but it is strange how people seem to carry on regardless.'

Firefighters deal with cases like this in quite frightening numbers. 'When people see something like that man's death and the way he died, it does give them a quick snapshot of the world you can find yourself in when you work on the ambulance. It's a world that not many people could understand and in many ways that's why we do this job,' says Alan Finn.

Paramedics and firecrews encounter death predominantly in natural circumstances. Contrary to hysteria and hype, we don't live in a brutally violent society. However, they also see death in gruesome, sad or humiliating circumstances. 'It's very rare that a case where death occurs affects me personally, I'd say it'd be the same for most crews,' says Finn. 'That said, you do become conscious of seeing death in the most mundane fashion some times. For instance, people who have simply pulled over to the side of their road in cars and passed away. Where was that person going? What happened? Suicide is different, we see that too, but when death simply happens it can trigger reactions in people which can be hard to define.' He also tells a story where he and his crew had to watch a father discover that his son was dead. 'It was on a building site in south Dublin, we were called out from Dun Laoghaire. We arrived at the site and in a deep trench was the body of a young man, face down. Myself and the s/o got down there as quickly as possible to see what we could do, but it was quite obvious the man was dead. A ten-tonne ductile steel pipe had broken free from a harness on a crane and struck him on the head. It would have killed him instantly. We got him out of the trench and were putting him in the ambulance, the other guys on the site told us that he was only twenty-five and had been working there for just a few weeks. They also told us he was the son of the site foreman. We could see the foreman walking up towards the site, to see what the commotion was all about, no idea he was about to discover that his son was dead.'

The Ambulance

In Dublin firefighters are also trained as pre-hospital care professionals and the ambulance is without a doubt the busiest element within Dublin Fire Brigade. Operating since 1898, it is a unique facet of the brigade, one of which they are very proud. It is the only brigade in the country which operates an emergency ambulance service in addition to their firefighting and rescue duties (indeed there are precious few brigades worldwide who provide a similar level of service). In the earliest days, firefighters were trained in first aid and operated horse-drawn ambulances. In the first year of its service, the ambulance was called upon 537 times. That number has been on the increase to this day. While serious building fires are on the decrease, the number of ambulance calls is on a vertiginous climb. For example, in the year 2000 alone, Dublin Fire Brigade responded to well over 140,000 emergency calls, over 94,000 of these ambulance related. The Dublin Fire Brigade is essentially contracted by the Eastern Region Health Authority (ERHA) to provide this service, which they do with a small fleet of just eleven ambulances. The ERHA has its own ambulance fleet, which does answer emergency calls, but only if a DFB one is not available. Essentially, dialling 999 for an ambulance will result in one of the brigade's ambulances responding. Each ambulance is crewed by two Paramedics. Until recently they were referred to as Emergency Medical Technicians (EMTs). These paramedics are also firefighters, but every firefighter has ambulance duty as part of his/her roster on a rotating basis. It is without doubt the busiest part of their duties.

DFB's ambulances are rarely used for routine ambulance transfers, they are dedicated almost solely to emergency work. As a result, they come face to face with very extreme cases. Stabbings, shootings, beatings, drug abuse, alcohol abuse, road traffic accidents are all familiar scenes for the brigade's paramedic crews. Ambulance work is done only by personnel of firefighter rank: once a firefighter is promoted to a higher rank, he works solely on the fire appliance or other emergency tender. Working on the ambulance is an intense, visceral work. 'You see it all, the good, the bad and the very ugly,' in the words of one paramedic.

Like on the fire appliance, every ambulance callout is a complete original. Some years ago, John Lynch, Alan Doyle and a new recruit from B Watch in Dolphin's Barn were responding to an ambulance call in Clonee, Co. Meath. 'When we got the call, we knew it was an urgent maternity case. We saw it on the incident summary as we were going out the door. However, we didn't know it was quite how urgent it would be,' remembers John Lynch. When the ambulance arrived, the patient, who was a non-national with little or no English, was already two days' overdue. This was not her first pregnancy and first-hand experience amongst paramedics has shown that after the first birth the process of delivering is, generally speaking, less complicated. 'We got her into the ambulance and started off towards the Rotunda, which is where she was attending,' remembers Alan Doyle. Lynch, who was driving the ambulance alongside the recruit, remembers asking Doyle what the condition of the patient was. In the rear of the vehicle, the woman had in fact passed the point of dilation and had gone into the final stages of labour. 'I remember Alan just saying to me, "John, I think you better pull over."'

With the ambulance stopped both men prepared to deliver the baby, hopefully with no complications. However, during the delivery, the paramedics saw that the umbilical cord had become entwined around the neck of the baby. The mother and the baby were now in a highly complex and potentially dangerous situation. 'There are some things that you can't train for,' says Lynch, 'I remember thinking that with this situation. When we were doing our emergency medical training with the brigade, eventualities such as this were done with a mannequin in a training room with other students and instructors. Learning the mechanics of situations is one thing, but actually dealing with a live situation where you're in the back of an ambulance at the side of a road with one other trained paramedic, a newborn baby in difficulty and a mother in discomfort who you're barely able to communicate with is something else entirely.' In this case, quick action was needed, to safely remove an umbilical cord from around the neck of a baby.

One of Doyle and Lynch's colleagues in Dolphin's Barn, Keith Carolan, remarks that it is a very difficult thing to do in the correct manner. 'Birth complications are highly complex situations,' Carolan says. 'Firefighters tend not to make a major issue out of a situation and this was one that they handled well. But what they did was no straightforward matter and they deserve a clap on the back for doing it in the way they did.' That day both paramedics, working together, carefully disengaged the child from a dangerous situation. 'We weren't the first or the last guys to do it,' says Doyle, 'but it was good that day to see a baby boy arriving safely. When he was delivered, we wrapped him in a foil thermal blanket and got both mother and baby to the Rotunda.' Later that evening, they received a call from the hospital to thank them and congratulate the two paramedics on what they had done.

An incredibly high amount of the callouts which stretch the overburdened DFB ambulance service are due to alcohol abuse. 'It's very frequent,' said one paramedic. 'For instance, on our last callout, we had an old man blind drunk and asleep on the pavement. He just asked us for a blanket and he'd be fine. Of course, we had to do a bit more then that.' Familiarity with the rhythms of the city mean that they know, purely by what's going on in Dublin on a certain day that they will be more busy then usual. Certain days are notorious, St Patrick's Day, Hallowe'en, the run-up to Christmas and New Year's Eve, to name but a few. 'To be honest, it doesn't even take major events for us to notice a fairly heavy increase in the workload,' he adds. 'It could be a simple thing like a big game on the early, which means people drinking heavily from early in the day until late at night. It's amazing the amount of people in football shirts you see in Accident & Emergency after a big game.'

Also, the carnage in pubs can be amplified in horrific fashion on the roads. 'You see some nasty things at RTAS, no doubt about it,' says FF Paul Green. 'One of the worst things can be crowds of people around a scene, making it difficult for you to do your work. One senior firefighter said to me once that you can tell a really bad RTA apart, because there is nobody around it. If it's really nasty, people don't stop and stare because they just don't want to see something really gruesome.'

Whether it be a fire, medical emergency, road accident or simply a hoax call to attack them, firefighters who face potential mayhem at an instant's notice can have a relaxed, almost detached demeanour when not on a callout. A busy station in Dublin can have an environment which would be the envy of any Human Resources manager. On one visit I make on a wintry Monday night, the entire crew on D Watch are sitting around a long pock-marked table, painstakingly counting out the proceeds of a DFB collection in aid of the victims of the Christmas 2004 Tsunami. The mood is one of laconic laughter, droll humour and persistent, good-natured, but robust, slagging. The station officer, Greg O'Dwyer, is eager for the personnel to tell me their stories. Some shift awkwardly, some get up to make a cup of tea, some point to those beside them or opposite them: 'He has plenty of stories, ask him . . .' They can be very hard to pin down. Greg O'Dwyer explains that it is in the nature of the firefighter to look forward, not backwards. 'As you will hear again and again, people in this job have plenty of things that they could tell you, but they are not in the habit of storing events and stories from callouts for future reference. One of the unspoken habits is that we don't take the job home with us, that's just the way it is. Some of the stuff we deal with, if you tried to remember it all, could drive you mad.' The raw material for the many fascinating stories within the fire service often originates with scant substance and the firefighter almost universally distancing him or herself from any credit for the work that was done. Modesty is a prized trait and in some ways could be seen as a shield for those who are uncomfortable in relating their achievements. It could be analysed and interpreted in any number of ways but that's just the way it is and, it seems, the way it has always been.

The 'brotherhood' of the service also extends beyond a firefighter's working years. On the first Thursday night of every month, in the Metro pub on Parnell Street, Dublin Fire Brigade's retired firefighters gather together. It's a close and raucous affair, no shortage of pints, banter and stories. Some are recent retirees, others left the brigade decades ago. They all enjoy an exceptionally close bond. According to Bertie Horgan who left the brigade over thirty-five years ago, 'There are very few men in this room who I would not trust with my life.'

These men are part of a thinning generation which I've heard today's firefighters describe as 'real firemen'. Some of the biggest fires in the city have been fought by these men. The tragedy of the Noyeks blaze on Parnell Street in 1972, the devastation of the 1974 Dublin and Monaghan bombings and the infamous Stardust inferno in 1981 to name but a few. However, when this generation of retired firefighters reminisce about their time in the service it is often the smaller memories, the sometimes blackly humorous and tragic details, which come to the fore.

Gerry Bell served with the brigade from 1973 to 2003, almost an epoch in terms of how the city has developed and changed. He remembers the surreal sights of his first callout on the ambulance. 'It was on the Clontarf Road, a lady was out cutting the grass, when she dropped dead and when we arrived her hand was still clutching the lawnmower, the motor still running.' Gerry served in the now-defunct Buckingham Street station from 1973 until 1984. While he has fond memories of the men he served with, to describe the station as basic in terms of facilities would be something of a compliment. 'The first twenty-four hours I spent in that place had me thinking that I might have made a serious mistake signing up to the brigade. Thirty-two firefighters worked there overall and there were no showers, just one nice big rusty bath for all of us.' Then of course there was the area itself. 'You might think the Buckingham Street area is a bit rough now, but my God you should have seen it back then. Madness. It was the toughest area in the city without question. There was Corporation Street, Foley Street, Sean MacDermott Street, Railway Street and the entire East Wall area. Complete nightmare at times, frightening.'

John Small, who worked with Gerry in Buckingham Street, concurs. 'Occasionally you hear stories of the rare old times, full of decent people. Well, we get attacked now in rough areas and by God we got attacked back then.' Gerry Bell also remembers a callout to St Anne's Park in 1975, where Garda Michael Reynolds had been fatally shot while pursuing criminals. 'He had been shot in the head, he was dead by the time we got there, terrible case.'

The majority of firefighters are family men, as proud to see their children grow and develop in life like anyone else. Cases which involve harm or death to children are always difficult for them to forget. 'The worst is cot deaths,' says Gerry. 'They always stay with me. I remember one Christmas day morning we went to a house where a baby had died, just passed away in its sleep. The father wouldn't let the child go, just kept holding it. Horrible scenes, but after that I had to come home and sit around the table and have a Christmas dinner with my family, looking at my own children and just being grateful for what I have.'

Tony Daly, who served 27 years as a firefighter in the city has similar memories of broken families. 'I remember a tragic ambulance callout in north Dublin where a gate had fallen on a child's head, only a young boy. We were rushing the victim to hospital and I was looking at his face, realising he was already dead. His father kept asking me was he OK and waiting for me to reassure him. The horror of being in that sort of situation certainly stays with you. Like everything else in the brigade you got to accept it as part and parcel of the work but you never got used to it, especially when a child was involved.'

Tony remembers the precious cups of tea back in the station after callouts like that, where what happened would immediately be brought out in the open and laid bare and discussed, the firefighter's way of dealing with trauma and tragedy. 'We didn't have the formal emotional support structures back then that the Brigade has now, it was more a case of sitting and talking with the other lads. That's something we used to do a lot. We'd eat together and discuss things, it definitely helped because you could air any problems you might have about something that happened. Other lads there could be in the same boat and once you talked about it, it never seemed as bad.' Ray Leavy, another recent retiree, joined the DFB in 1976 and like Gerry and Tony one of the callouts which sticks in his mind was tragedy involving a child. 'An incident on Thorncastle Street, near Ringsend, was one I particularly remember. A truck carrying cowhides capsized on the road and killed a four-year-old child. Very difficult to deal with.'

Of course, the job can present incredible, life-affirming memories too. 'It's a great feeling when you can save somebody, make that critical intervention,' says Tony Daly. 'If you manage to resuscitate somebody or stop the bleeding or just really make a difference, it's fantastic.'

One of the longest serving members in the brigade was Liam 'Nobby' Clarke. At the time of his retirement he was arguably the oldest serving member of the DFB. He joined the brigade in 1963 and only retired in the spring of 2004, breaking the 40-year barrier beyond which very few go. Like many of the other retired members, when 'Nobby' talked of joining the brigade it wasn't because of a strong vocational desire, it was because it was a steady job at a time when such things weren't in abundance. 'My father, Billy Clarke, was also attached to the brigade. I remember he told me to get into a job where there was a uniform because there you would have a job for life. I have to say the man was right.'

Such was Nobby's time in the brigade, mostly served at Phibsboro station, that his face could almost have ingrained itself onto the walls. He was widely acknowledged as one of the DFB's most engaging characters who in addition to being a legendary firefighter, was also a participant in some of the brigade's most legendary practical jokes. The 'hand' which stayed in Nobby's locker until the day he left the brigade, 'that came from a fire in a clothes shop. I took the two hands off one of mannequins, I picked them up and thought they might come in useful. I was with my old friend Miley Keogh at the time, who has since passed on. So, from time to time I'd stick the hand in my pocket when I was going out on the ambulance. We'd be in the old Richmond hospital on James's Street with the hand sticking out of my pocket, just to see the look on people's faces when they first saw it. Of course Miley would add a bit of drama to the moment by saying, 'My God, Nobby, why didn't you leave that hand here after the last accident.'

As well as cracking jokes, he was often the butt of them. 'I could be sleeping and then the bells would go off, so we'd turn out on the appliance and later when I get back to the station someone would have moved my bed, I've lost count of the amount of times that's happened.' In addition to the sheer madness of work in a fire station, the camaraderie and the laughter which kept him in the job for forty years, Nobby also has a wealth of other memories, some dark and some dramatic. 'I've seen so much and of course I've forgotten a lot. I've seen a lot of death, I know that. Going into a house fire and finding some poor soul's body in the dark, that's happened a lot. Of course I've also seen life, which is fantastic. Between houses, lifts and of course in the back of an ambulance I've delivered about twelve babies into this world.'

He's also seen times of terrible turmoil in the city. One which sticks in his mind and was the fire at the British Embassy on Merrion Road, which was burnt to the ground after the Bloody Sunday massacre in Derry in 1972. 'I remember that night, the city was seething with anger. The embassy was ablaze with thousands of people watching it burn. The atmosphere was very volatile, one of the darkest days of the

Seventies. We turned out to the fire, under the great D/O, Mick Delaney. We were on our way from Tara Street station and just as we got to the fire I remember an individual coming over to the fire appliance as we were stopped. Mick Delaney was a tough no-nonsense individual but this guy, whoever he was, said something very serious to him. To this day I personally maintain that this individual opened his coat and showed a gun. The meaning was clear, they didn't want us to do anything about the building. I remember Mick just looked at the driver and told him to head back to the station. That was quite a night.' Other milestones in Dublin's history which Nobby witnessed were the destruction of Nelson's pillar by the IRA in 1966 and the night when one of the angels on the base of the O'Connell monument was attacked: 'Not many people remember that, but I turned out to it.'

Like so many in the Dublin Brigade he has his personal memories of Stardust. 'That fire fundamentally changed the brigade. I was on duty that night. A colleague of mine, Jimmy McDermott, since deceased, had one son and two daughters killed in that fire.' Although he didn't know it at the time, Nobby discovered the body of his colleague's son behind the counter in the charred interior of the Stardust. One more tiny tragic story from a night the city's firefighters will never forget.

Leaving the Brigade can be traumatic, like leaving a close family with whom you have eaten, slept and worked with for decades. Tony Daly says there are mixed feelings every time he walks past Kilbarrack station, where he worked for twenty-seven years. 'There are times that I pass the station and I look at it and think I'm glad I'm not in there. The camaraderie is something I definitely miss, but a lot of the lads I knew are involved in other things with me so you do get to keep in touch. I also have a son in the service, which keeps the connection alive.'

For Gerry Bell, it's time to put on the slippers. Looking at the bronze retirement statuette on his mantelpiece, he says, 'After twenty-nine years in the service it wouldn't be natural to just walk away. I do miss the men I worked with and the ambulance in particular. I served my time, though, and I'm exceptionally proud of my career in Dublin Fire Brigade.' Ray Leavy, who served the capital for twenty-seven years, is typically low key when he describes his career: 'I suppose I would say that I never changed the world or anything, but I know that I always did my best.'

Water Rescue

The River Liffey is one of the major killers in Dublin and in the experience of Dublin Fire Brigade, it easily accounts for more deaths in the capital then any other specific location. Thus, water awareness and water rescue has become an area of high priority for the brigade in recent times reflecting the prominent and growing role that this type of rescue work plays in the life of a firefighter. While the number of overall callouts for the brigade relating to water rescue is only in the region of 5 per cent, for some stations such as HQ on Tara Street it is a very regular occurrence. Here the percentage of calls where SRT — Swiftwater Rescue Technique — is required is much higher, indeed HQ would respond to over 95 per cent of water-related callouts for the entire brigade. It's a different discipline and so, a different set of rules apply.

Prior to the development of a rigorous water rescue programme in 2001, the fashion in which these potentially treacherous operations were undertaken was far more ad hoc. 'It's state of the art now and constantly improving, but before we had this training and this new equipment we just did it any way we could,' says HQ S/O Greg O'Dwyer. 'Some very effective rescues were carried out using rudimentary equipment, basic knowledge and common sense. We would go in with a rope around our waist for example, which we now know is highly dangerous.'

While it may seem like the simplest and most effective way of keeping a rope on someone in the water, the death of a firefighter in Manchester in 2000 highlighted exactly how dangerous it could be. Ropes which aren't designed for SRT aren't water resistant and can become quite heavy in the water. In addition, when the rescuer is tied from behind and if they should become trapped by an entangled rope it can be most difficult for them to free themselves. 'It was only through the water awareness and subsequent SRT training that we realised how dangerous the methods we were employing actually were,' admits Greg O'Dwyer. 'The equipment we have now is designed professionally for SRT, down to the ropes which are made from lightweight polyurethane and float in the water. The rope is now connected to the rear of the SRT harness but it can be disconnected via a quick release hook at the front of the harness, which means the rescuer is always in control.' In the next year, over half of the city's operational firefighters will be trained in this burgeoning discipline and it's not just in the capital, SRT is also underway in the full-time brigades in Cork, Limerick and Galway to date.

Despite all the equipment though, SRT and the situations into which it can put a firefighter can still be hazardous. 'Like anything else we do, an SRT firefighter is deployed in a certain way depending on the circumstances in which we find ourselves. Everything, be it a fire or a water rescue, is subjected to a dynamic risk assessment,' adds O'Dwyer. This is a common practice in Irish firefighting and rescue and essentially means that a varying set of safety protocols are put in place to make sure that the operation is done properly and safely. 'If we're going to rescue someone, be it from a fire or the depths of the Liffey, it's part of our job to ensure that we don't potentially end up with two casualties, one of them being a firefighter. It's just not going to happen if we can possibly avoid it, that's why we have such a good safety record.' That's not caution to today's firefighter, it's common sense. 'There are very few things that we won't be able to handle,' adds O'Dwyer, 'even if we haven't encountered it before, we'll make a fairly good effort at dealing with it anyway.'

With rivers, canals and Dublin bay itself, the capital is well stocked with water-rescue challenges and just like firefighting, when the bell goes, the form book goes out the window and the old mantra of 'every callout is unique' is as true as ever.

Most callouts involving water are undramatic and resolved relatively quickly, with firefighters effecting a quick rescue or a grim recovery exercise. However, there are times when a series of surreal and potentially tragic events conspire to present a callout which is anything but low profile. Shay Rowe, based in HQ on Tara Street, remembers such an incident, which occurred on 23 July 2002. A large heavy goods vehicle (HGV) was involved in a near-miss with a bread van on Tara Street bridge. The drastic action required of the truck driver to avoid collision led to the vehicle smashing through the concrete barrier on the bridge and plunging into the water below. The sight which greeted the general public as dawn rose was that of a truck with its cab practically submerged in the water and its trailer high in the air, the back wheels still on the bridge.

Once the alarm had been raised, DFB crews arrived very quickly as the station is a matter of moments away. 'The first thing we did when we got there was raise voice contact with the driver, to see if he was still with us,' says Shay Rowe. 'What had actually caused the accident was no concern of ours at all. All we wanted to do was get him out and make sure he was OK.'

At any scene like this, speed is of the essence but so is having the right plan. 'We kept in voice contact with him and told him not to move. The first plan was to consider whether we could put a ladder in the water and retrieve him that way but with the angle that the truck was at, that wouldn't have worked.' It quickly became obvious that one of the firefighters would have to enter the fast-flowing water and help the driver escape. 'Thankfully, the water was quite shallow at the time. If this had happened in winter, it could very well have been a body recovery exercise.'

While plans to insert someone into the water continued, a second unit of the Brigade arrived on scene. 'The bread van involved in the crash started to go on fire so we needed a second unit to deal with that.' Meanwhile Shay Rowe started his descent into the Liffey via one of the ladders and attached to a safety line. This rescue took place before the SRT revolution in Dublin Fire Brigade and as a result used different practices to those currently employed. 'When I looked down, I thought there was quite a bit of water but, as I said earlier, when I got down there it was very shallow. That's OK from a water safety point of view but to put it mildly there is a lot of "unsavoury" stuff on the bottom of that river.' After negotiating his way through a small maze of traffic cones, shopping trolleys and other urban detritus, he was finally able to breast stroke to the cab of the truck. 'When I got to the cab, the driver was sitting there surrounded by water up to his waist. Thankfully, and fair play to him, he wasn't panicked and was just watching the rescue unfold.'

Between the two of them they managed to get the window down and Rowe was just about to help the driver out when he asked could he take something with him. 'They always say in an emergency to just drop everything and leave but working in fire and rescue you get used to people asking you can they take stuff with them, it's natural I suppose, but normally it's stuff that they might really need like a set of keys, a wallet or a mobile phone. This guy looked at me with a straight face and asked could he take his Playstation with him!' Dragging someone through a window, fitting them with a lifejacket and getting them safely back to the quayside was difficult enough without the added burden of a sizeable piece of hardware technology. 'I though the Playstation was a bit much to be honest, so we just got

Dublin firefighters at the scene of a massive blaze at a print factory in Cabra in 2002.

(Picture: Colin Keegan/Collins. Courtesy of *Firecall* Magazine)

A firefighter encounters a flashover as he enters a training simulator designed to mimic a bedroom engulfed in flames at the Fire Training Centre on Treasure Island in San Francisco. The Irish Fire Service provides a similar facility for in-service firefighters and new recruits.

(Empics)

Dublin Fire Brigade SRT (Swiftwater Rescue Technique) team during a training exercise. The use of this technique is growing rapidly in the capital and elsewhere in the country.

(Picture: Ron McGarry. Courtesy of *Firecall* Magazine)

Firefighters performing a structural firefighting exercise as part of their graduation ceremony at the Dublin Fire Brigade training centre, the O'Brien Institute.

(Courtesy of *Firecall* Magazine)

A Dublin firefighter undergoing marine firefighting training aboard a simulator.
(Courtesy of *Firecall* Magazine)

Marine Emergency Response Unit firefighters being winched aboard a vessel by a coastguard helicopter as part of a training exercise.

(Picture: Peter Shaughnessy)

Gerry Bell, a forty-year veteran of the brigade, who retired in 2003.
(Picture: Ron McGarry. Courtesy of *Firecall* Magazine)

The warehouse in west Dublin where the late Mick Loughlin and Noel Kelly faced a life-threatening flashover.

(Courtesy of *Firecall* Magazine)

Firefighters survey the scene after a truck ploughed through the wall of Tara Street bridge into the Liffey in July 2002. The driver was rescued unharmed.

(Courtesy of *Firecall* Magazine)

Mayo firefighters face the devastation at the Cow Comforts Factory in Belmullet, Co. Mayo, August 1999.

(Picture: James Connolly/GreenGraph. Courtesy of *Firecall* Magazine)

Seamus Murphy, Chief Fire Officer, Mayo Fire Service.

(Courtesy of *Firecall* Magazine)

Powerscourt House in Enniskerry, Co. Wicklow is reopened in 1997. The House had been derelict since 1974 when it was tragically gutted by fire just after extensive refurbishments had been carried out by its owners.

(Picture: Eamonn Farrell/Photocall Ireland)

The MV *Betelgeuse* on fire at Whiddy Island oil terminal, 8 January 1979.

(Picture: Richard Mills/Irish Examiner)

him out of there without it.' At this stage a fair number of the city's early risers were watching the drama unfold but to the initiated this was as straightforward as a water rescue gets. 'Considering the surreal situation in which it happened, it all went very smoothly; they're not all like that.'

Some of them are a lot tenser. 'I've attended quite a number of serious incidents on the Liffey in the three years I've been here,' Rowe said in 2002. 'Sadly, some of them have been fatal because there is still a lot of underestimation when it comes to that river and of course, there are always tragic accidents and events.' In May 2002, the crew from D Watch in HQ were called out to an incident near the Custom House. 'When we arrived, there was a respectable-looking woman in quite a state, shouting and pointing into the water.' Apparently, there was a man in the water. How he got there wasn't important, just the fact that he was there. It was pitch black, so the firefighters had no point of reference until the woman pointed towards a plastic cup which was bobbing in the waters of the Liffey. 'If you really strained your eyes you could see what looked like a human elbow breaking the water,' adds Shay. 'I had the harness and line on so I went over the wall and into the water, it was only about a six-foot drop. The drinks cup was playing a vital part in how we did this as it was our only consistently visible point of reference. The other lads there such as Keith "Cabbage" Pearce were directing me to where the guy might be by just bellowing instructions as to how close I was to this plastic cup.'

When he finally reached this nondescript cup, he could see nothing, just inky black, cold water. 'At that stage, I just moved my foot around beneath the water to see if I could feel anything and I hit something which moved. It was him.' Diving into the water he pulled the man clear and signalled to the crew to pull him back in. 'As I was hauling him back up the water I was pretty sure he was dead but nothing is certain in this job.' Laying the victim on the ground the crew went to work. 'We cleared his throat to assist him if he was able to breathe and gave him several chest compressions and some artificial respiration. Whatever we did, we did it well because he started to come round and to vomit out a significant amount of the Liffey. We then did some more compressions and he managed to clear his throat. At this stage the ambulance was there so he was whisked away to hospital. That was one of those cases where everything worked out, we did our job as quickly as we could and the guy was very, very lucky. Blessed in fact.'

Greg O'Dwyer remembers an off-duty firefighter on his way to work when he saw someone jump into the Liffey, not a rare occurrence. 'I remember it was close to Christmas so the water would be particularly cold, things moved very quickly.' Although the firefighter in question was bereft of his equipment and colleagues, he did have his training and within moments was scaling the metal railings of the bridge and leaning down over the dark, fast flowing river to intercept the man floating down towards him. Meanwhile, DFB had been informed and dispatched an appliance to deal with it. 'It was rush-hour traffic at Christmas so even though we had the lights flashing and the sirens blazing it was impossible to get past O'Connell bridge. Back at the Ha'penny bridge, the firefighter had managed to secure a grip on the casualty but he was still in the water, which was flowing quite fast, and the firefighter couldn't pull himself free so there were two people in need of help although, the rescuer's swift and selfless action surely saved the man's life. Meanwhile, much to the surprise of Christmas shoppers, firefighters in SRT gear were running to the aid of their colleague, complete with a 10.5-metre ladder. 'When we got there we managed to get the ladder down into the water so the firefighter could get onto it and pull the casualty up with him. It was one of those things that you actually rehearse in SRT drills and it worked wonderfully well on this occasion.'

Firefighters are well used to using every piece of kit in their armoury to effect a rescue and indeed, under the Fire Services Act, are entitled to use anything else that might save a life. 'We had a case where Tom Butler, one of our best SRT guys, commandeered a boat to rescue a casualty. The casualty had fallen off, or jumped off, one of the Liffey quay walls. Since the water was quite shallow, Butler had to assume that he may have suffered spinal injuries.' Firefighters could have got to him by using the ladder but could also have done more damage to him by hauling up a vertical ladder. 'So Tom saw a boat at the Liffey Voyager berth and commandeered it, securing the casualty in the boat and taking it down to Butt Bridge where we could immobilise the man properly on a spinal board and cervical collar and get him to hospital. That's a good example of getting the job done and using whatever resources you can.'

DFB are taking SRT so seriously now though that they have obtained their own boat for rescue purposes. Of course, like so many other parts of their job, the hard-earned skills of firefighters are occasionally put to use as a result of what could only be described as human stupidity. Such stupidity is often the result of drunkenness, with often undignified and dangerous results. Take for example what happened at a bridge over the river Dodder on a recent New Year's night, related by a firefighter. 'This is the sort of thing where you don't know whether to laugh or cry, such is the stupidity. A girl was obviously on her way home from a New Year's party, couldn't get a taxi, the usual story. Anyway, there was a fair bit of drink onboard shall we say. So this girl suddenly gets the call of nature and decides that it's a great idea to alleviate this problem over the Dodder bridge. Now as we know, balance and alcohol don't make a great combination.' A fact which this lady was to discover. 'We got the call from her friends who had seen what happened.' In an attempt to use the bridge as a makeshift toilet she had lost her balance and fallen into the river below. 'This resulted in quite a dangerous situation as we had a casualty unconscious after falling a considerable distance, added to the fact that she was in a rather undignified state of undress.' Treating the situation as a potential spinal injury, firefighters extricated her from the mire of the Dodder shallows and she was taken to hospital where, rather amazingly, she wasn't injured. 'I remember coming back from another callout and we saw her walking down the road again, obviously the hospital staff hadn't given her much sympathy once they realised she wasn't hurt. One of the weirder water rescue incidents.'

It is often the shallows of the capital's waters that present the biggest threats. 'We've had numerous incidents of people stuck in shallows, embedded in mud. It's not the mud that's dangerous so much as what's in the mud,' says Greg O'Dwyer. Buried in there are the results of industry, pollution and dumping going back generations. For instance, substances such as mercury have been identified in the shallows and mudbanks of Dublin Bay. Thus, in addition to the challenges of extricating casualties, the firefighters have the whole issue of contamination to deal with. The drysuit which the firefighters use when performing SRT is designed to protect them from this hazardous environment. 'We also have decontamination showers deployed in these sort of environments. In addition to the threat of contaminants there is also disease to contend with.' The most feared of these diseases by firefighters is without doubt Weil's disease, carried predominantly by rats which frequent the waterways of the capital. The disease is extremely virulent and potentially lethal. 'We have had one extremely serious case of Weil's disease within our SRT squads over recent years, so we take no chances,' adds O'Dwyer. When SRT crews return from duty they have a mandatory decontamination shower with specially designed antibacterial gel. The PPE itself is hosed down with an antibacterial solution. As with everything else, safety comes first.

The sharp end of water rescue occurs when elements such as fire and water combine. Ask any firefighting expert and they will confirm that a fire aboard a vessel at sea is amongst the most hazardous situations which a firefighter can encounter. Fighting such a fire requires a mix of water-rescue skills in addition to firefighting.

Being an island, it was recognised in the mid-Nineties that Ireland needed to have a very robust response to extreme situations in our territorial waters. Essentially, we needed an elite fire and rescue unit, specially trained and prepared for these eventualities. The result is the Marine Emergency Response (MER). This is a group of firefighters from DFB, based in the capital, prepared to respond to any incident nationwide which requires their specialist training and knowledge. All the firefighters in the MER are operational personnel, and respond to alerts by mustering with their MER colleagues if and when the situation requires. The unit is a stand-alone asset of the Department of the Marine and they can be airlifted by helicopter at the direction of the coastguard, with the agreement of Dublin's CFO, from Dublin airport to respond to a situation anywhere in Ireland's territorial waters. The situations can be extreme and as a result, so is the training.

To date, their full array of skills has been called upon only once. In 2002, the Baily lighthouse became flooded with carbon dioxide, a highly hazardous situation which could have led to a major explosion. MER was airlifted in to vent the building. However, while the callouts have not been many, the challenges are there. Acting District Officer Gerry McMahon is one of MER's instrumental officers. 'We have a lot of Russian and Spanish fish factory ships in operation, particularly off the west coast. A fire on one of them would be a very serious situation.'

In addition, Ireland's extremely busy ferry routes out of Dublin,

Rosslare and Cork have some of the world's largest passenger ships. It is being able to respond to a fire on such a ship, such as the immense Irish Ferries vessel MV *Ulysses* or the Stena *Adventurer* which is one of MER's key tasks. Refresher courses are held annually on these ships or similar vessels to remind members of MER of exactly how extreme the challenges can be. The training has become renowned, or indeed notorious, for being extremely harsh. 'The refresher courses are a week long and include an exercise where personnel are winched onto a ship at sea to carry out a BA exercise in the machine room,' adds Gerry McMahon. Without the ship actually being on fire, this is the closest that instructors can place firefighters to the real thing. 'We clip clouded Perspex onto their visors so they can see light but little else. So there's very little visibility and it's very hot. On the Stena *Adventurer* they had to climb down escape shafts into the machine room.' A ship's machine room is an extremely cramped and humid area, a completely alien environment through which firefighters must navigate. MER personnel experience the simulated effect of having to find and extinguish a fire in tough conditions, smoky and dark with tight spaces. Safety is, of course, paramount though, and there are two instructors on each exercise exactly for this reason.

In addition to the physical challenges, there is also a psychological perspective to an extreme scenario like this. Working on an unpredictable vessel like a ship is different from any building and MER personnel say that they do notice a difference in their psychology and the psychology of their team. They arrive by helicopter, winched on deck with a heavy kit bag consisting of PPE, BA and firefighting equipment, so from the start of the exercise the environment verges on the extreme.

The ship simulation module is located in the DFB's training centre, the O'Brien Institute, and is a stand-alone, steel, four-storey simulator. It is designed to replicate the obstacles MER personnel would face on a seagoing vessel, including steep and narrow gangways and passages, vertical hatches and crawlspaces and confined machinery rooms. Inside the simulator, firefighters are subjected to temperatures of 60 degrees centigrade. This is merely a warning temperature. On a fire at sea, temperatures of 160 degrees are common. According to the instructors, the point of MER training is to make everything as real as possible and to leave nothing to chance. 'The course has a very steep learning curve,' says Terry McMahon, 'there's certainly an element of terror involved, but I prefer someone with a bit of fear.'

In addition to ship firefighting and sea survival, MER personnel are also trained in helicopter operations and forced ditching. Indeed, prowess in aerial situations is a vitally important part of the training and prior to selection the recruits must pass a flight crew medical. The helicopter ditching exercise is carried out at the Fleetwood Offshore Survival Centre near Blackpool in the UK, a facility which has a helicopter simulator and a tank which is forty feet deep. The facility can also emulate conditions such as poor light and fog.

The training comprises several exercises. In the first, the simulator hits the water and the crew must escape in a liferaft. In the second, the simulator hits the water but also rolls 180 degrees and the crew must escape. The third exercise has the simulator spinning a full 360 degrees in the water before the crew can evacuate. The second and third exercises are done in total darkness with the exception of a single fluorescent light for the benefit of the instructors and the safety divers in the water tank.

Colm Murphy, one of the DFB's foremost water rescue exponents, recently became an instructor for MER and remembers that his training to become part of this elite unit was the toughest of his time in the brigade. 'It was definitely the toughest couple of weeks. You're put into environments as severe as they can be. It's not for everybody, you need cool heads and strong bodies and there were moments when I wondered what I was doing there.' According to those who survive and pass the training though, the reason MER is so popular is because it's so tough; to meet the challenge at this level is an adrenaline rush in itself. To pass you must be fit and undergo a full blood, cardio and respiratory exam. To join MER you must also have five years operational experience under your belt.

Major exercises and constant training are part of the life if you're with the unit. Some of these exercises are run in conjunction with UK brigades, with whom the Irish firefighters have a great rapport and understand each others responsibilities in responding to incidents. On a recent training mission the MER personnel, in conjunction with their counterparts in the North Wales Fire Service, were winched onto a ship on the Irish sea in the midst of force-ten gales, with ten-metre-high waves buffeting the vessel. The buzz of successfully carrying out an operation like this is exactly why the MER courses are oversubscribed. It brings firefighting and rescue to another level entirely and while job satisfaction is a bonus, the reason MER exists is because it is needed.

Close Call

On every callout to a fire, death can be waiting. For firefighter Paul McGurrell, one callout in early January 2006 was a very close call. It was a normal day shift on D Watch, Dublin Fire Brigade. Normally stationed at HQ, he had been moved out to Dolphin's Barn, station number 2, for the day. As day shifts go it had been fairly standard, nothing dramatic to report. In mid-afternoon, a call came through of a fire at a commercial premises on Cork Street in the south inner city. McGurrell, with the other men on his watch was on Delta 2-1, the station's first fire appliance.

When they arrived on scene, the fire was well underway, 'whipping it' as firefighters might say. But it was hard to get to, since it seemed to be mostly raging beneath the roof tiles in the attic space. Working with another firefighter, Darren Donovan, they got onto the roof of a neighbouring building to see if they could make any progress in suppressing the blaze. There was no joy there so the officer ordered them into the building, where they started hitting the flames through the ceiling above the blaze. 'We could see the fire in places but there wasn't any way of getting at the seat of it from where we were.' McGurrell had only been in the brigade for five years so like most sensible firefighters, deferred to experience from the more senior men. On scene that day was a Third Officer, a rank normally stationed in HQ but often present on the fireground if there is a major blaze. 'I remember Third Officer Hederman coming into the building and taking a good look at the burning ceiling above and how the flames were moving. He then told us that he didn't like the look of it and we were to move out. When you hear one of the senior guys doing that, you just do it. They know what they're talking about,' adds McGurrell.

The firefighters withdrew from the burning building and started battling the blaze from the outside. The fireground had become quite busy now, at least eight appliances in attendance. Hitting the flames with the hose from the outside, scarcely two minutes after they had withdrawn from the building, McGurrell remembers hearing a crash and the ceiling under which they had been standing inside collapse in a pile of flaming masonry and rubble. 'Never thought about it at the time of course, but that was exactly where we had been standing.' With the ceiling collapsed, the firefighters were able to make better progress with suppressing the flames but the structure itself wasn't deemed to be safe and all the firefighting would be done from the outside. With no persons reported inside, there was no sense in taking unnecessary risks.

McGurrell was doing what he was trained to do, targeting the flames with water, standing alongside Firefighter Paul Foley when he noticed the wall above him start to shake and move. 'I just saw it shaking and then I remember hearing a shout from the lads behind me.' The firefighters further back from the building could see the structure beginning to tremble and shake and, sensing impending tragedy, screamed at McGurrell to get out of there. 'I wasn't aware of putting myself in harm's way, I was where I was supposed to be but when I heard the roar and rumble from the wall I just turned and ran, flinging myself forward.' The wall at the rear of the building crumpled to the ground. McGurrell was lying face down, just yards from the smouldering pile of masonry. Exactly where he had been standing was now a large concrete plinth. Firefighter Paul Foley, twice McGurrell's size, reached down and pulled him to his feet, with the exception of getting a nasty knock on the leg he was fine. If he hadn't moved, he would likely have been killed. 'I was OK, I mean I was a little bit shaken when I saw where I had been standing but apart from my leg feeling dead, I was fine.'

Firefighters are loath to dramatise their work, it's not their way. Like other jobs, their role has risks but simply because they are firefighters they do not feel that it warrants more discussion then anything else. That said, Dublin Fire Brigade had come awfully close to a tragedy that day. 'You know, I never really thought about it like that. You just do things automatically when you're on a callout, you react to things quickly. That day I was lucky I did.' He was discharged for the rest of the shift to ensure his leg was OK, which it was apart from a decent-sized bruise. 'I just got back into it after that. I didn't really think about it again. There was certainly no post-traumatic stress for me; what happened was just part of fire and rescue work. I had no problems or concerns walking back into the station.'

Of course, that's not strictly true. Once the danger has passed, firefighters have their own way of remembering an incident and ensuring that some are never let forget it. Safety is key to everything in the modern fire service, but also important is the brigade's famous, or infamous sense of humour. 'I'm waiting for it to start doing the rounds alright, I know I'll get some stick out of it for sure. That's the way it is in here, once you're ok and it's not in bad taste then you can get a fairly good slagging. To be honest, knowing the firefighters' sense of humour, I'm a little surprised that nobody has put a brick in my locker yet!'

Fire and Human Behaviour

Trying to understand the mentality of someone who would willingly set a fire may be a difficult concept but when you consider that humans have always been fascinated with fire, it's not so hard to imagine. It is and has always been a quirk of the human condition that fire, that intangible and fascinating substance, sometimes arouses extreme and dangerous tendencies. Whether an arsonist does or doesn't deliberately set out to cause harm or death, the stark fact is that they are meddling with such an unpredictable element that tragedy is bound to occur in some instances.

So what makes arsonists tick? There is a Freudian theory in which Sigmund Freud linked arson to sexual motives. 'The warmth radiated by fire evokes the same kind of glow as accompanies the state of sexual excitement, and the form and motion of the flame suggests the phallus in action.' So he theorised that the setting of fires was some sort of sexual surrogate for frustrated men or women. More contemporary research however suggests that this motive is extremely rare amongst arsonists. According to another analyst, Maurice Gold, if this were the case there would be a lot more arson. 'It isn't just sexual tension,' he says, 'if this were the case, then no city in the world would be left standing.' To give Freud's theory some credit though, it has been established that arsonists do have trouble forming relationships, particularly with the opposite sex.

For firefighters sifting through the ruins of a burnt-out home, such academic theories have little place. In September 2000, firefighter Charles McAuley of Belfast Fire Service broke down in court as he recalled how the lifeless bodies of three young brothers were passed out to him after an arson attack on their home. He was one of the first firefighters to arrive at the home of the Quinn family in Ballymoney, County Antrim which had been subjected to an arson attack. When McAuley entered the blazing building he began to tackle the flames around the door. He then saw one of his colleagues coming down the stairs, swathed in smoke and carrying the body of a young child. He went on to tell the court that after he passed the small body to waiting paramedics his colleagues emerged with two more blackened victims of the fire. His voice cracked with emotion as he told Lord Justice McCollum that he believed there were others trapped within the blazing house. The firefighters continued their desperate search by crawling through the dense smoke and flames, feeling with their hands for any more survivors or victims. 'In spite of this tragedy I am extremely impressed by the speed of the reaction and the courage and dedication of the firefighters. They did their best to reduce the damage and could have done nothing more,' said the judge.

The damage had been caused by a 1.75-litre bottle of Bushmill's whiskey and a dangerous mind. The bottle, which had been utilised as a petrol bomb, was recovered amidst the debris of the house and a lighter was found nearby. A twenty-four-year-old man was sentenced to three life terms for the murder of the young boys. The attack was planned by the UVF and the house was a sectarian target.

One of the more high profile cases in the Republic in recent years was a savage attack in Tallaght garda station which led to the death of Sergeant Andrew Callanan. He lost his life trying to save the life of the arsonist who walked into the station at 4.45 am in July 2000 armed with petrol and flares. The man set the public office alight, where gardaí were working. He eventually escaped but was arrested soon afterwards. Sergeant Callanan suffered severe burn injuries as he tried to get the arsonist away from the flames, he was rushed to Tallaght hospital but died from his injuries shortly before 6 am. The savage, cowardly and chillingly premeditated attack shocked the nation and highlighted the irrational nature of the arsonist.

Some do it purely for the excitement which the lighting of fires gives them, others from a misplaced rage against society for some perceived injustice. Since lighting fires requires little money and no courage, it is an easy refuge for the weak. In 1996, John Paul Buck from Clonmel got it into his head that he would 'get one over on the gardaí' following the arrest of his brother on another matter. He began by setting fire to the community hall and over the next few weeks targeted cars all over Clonmel, causing an atmosphere of fear and menace. When he was eventually caught he was sentenced to seven years in prison, the judge branding him as 'bizarre, outrageous and arrogant'.

Firefighter Colin Murphy has been nearly ten years with D Watch in Finglas. 'When I first started here, in the early to mid-Nineties, the so-called Drugs War was in full swing.' This was when the residents of the city's more impoverished and drug-riddled areas began to revolt against the insidious and often blatant presence of drug dealers in their areas. 'It was a time when we saw quite a lot of scary stuff,' Colin remembers. 'Arson was widespread. If a drug pusher was forcibly removed from the area, retaliation often took the form of arson, though they often took their revenge on someone who may have had little or nothing to do with that particular drug dealer's eviction from the area.'

There was a particular arson tactic which the firefighters became all too familiar with at that time, particularly in callouts to Ballymun. 'What they would do is fill a hypodermic syringe with petrol or another highly flammable fuel. They would then inject that in through the letterbox of the particular flat. The flats in Ballymun all have short, straight hallways just inside the door. By injecting a jet of petrol through the letterbox, followed by a lighted match or piece of paper, they could get a trail of flame through the hallway that would easily take hold. Many flats were burned out in this way. Sometimes there were people in them at the time, sometimes there weren't. There was a vicious spate of them at the time though. One girl I remember in particular had just gone to collect her daughter from school, she wasn't from the area, but somebody had taken the notion that she was heavily involved in the anti-drugs campaigns. She returned home to find her little flat destroyed by fire. That's the way it was, random, indiscriminate and dangerous.'

The list goes on: Terry O'Connor from D Watch in Tallaght station can't get a fire which occurred in 1986 out of his head. 'A mother and her four children perished in that fire, which eventually turned into a murder investigation.' For his colleague, Gary Hayden, a tragic arson attack on two boys is one of the most chilling memories of his relatively short period in the brigade. 'I remember there was a little hut on a piece of waste ground, the two boys were playing in it. For some reason, someone set fire to the hut with the two children inside it. One of them escaped, but the other didn't. I distinctly remember crawling through the burnt-out shell of that hut, it was black with smoke, feeling with my hands to see if I could find the other lad. I eventually put my hand on something and I knew it was him, I couldn't see him but I knew I'd just put my hand on his dead body. That was a terrible night.'

Most cases where arson is involved require a psychiatric report which delves into the mind of the offender. The conviction rate in cases involving arson is quite low — American figures suggest as low as fifteen per cent — while arson is cited by the Americans as being the second leading cause of fire deaths and could count for as much as twenty-five per cent of all fires in that country. However, while the motives could have a psychological root, arson is frequently the result of plain and simple vandalism. There is strong anecdotal evidence that suggests that some young people in Newcastle in the UK are required to burn down a building before being accepted into a gang. Demographically, arsonists are traditionally male and have low socio-economic status. Alcohol and other substances have been found to play a major role in their actions. Arsonists generally seem to suffer from urges rather then one untreatable condition and there is quite a low rate of re-offending if the arsonist is caught and brought to justice. There was an even lower rate of re-offending in centuries past; in 11th-century England arson was punishable by death. This was re-evaluated in later centuries, though, and while the death penalty didn't apply, the offender had one hand and one foot chopped off and was sent into exile.

Today, there is a far more scientific approach in trying to decipher the hidden truths behind an arsonist's desire to set fires. Australia, with its continuing wildfires frequently caused by arson, has done extensive research in this field. The Australian Institute of Criminology has pointed out that while an arsonist's firesetting is frequently simply described as 'pyromania', this is inaccurate:

'Pyromania is an established psychiatric diagnosis in the *Diagnostic and Statistical Manual of Mental Disorders*. It falls into the category of 'impulse control disorders,' along with disorders such as kleptomania, intermittent explosive disorder and pathological gambling. These disorders as characterised by a failure to resist impulses, such as the impulse to light a fire.' The authors of the Australian report contend that the pyromaniac has a degree of fascination with fire that goes well beyond the curiosity frequently felt by a child. A dedicated pyromaniac does not feel remorse and will not be concerned with the potentially devastating effects of their actions. For the pyromaniac, the fire is not a means to an end but and end in itself. The Australians also contend that such an extreme level of pyromania is extremely rare. This report's main point is that you cannot dismiss all arsonists as pyromaniacs. There are many other influences, such as the sectarian one which led to the death of the three Quinn boys.

Most arsonists are totally in control in their decision to strike a match and set in motion a lethal chain of events. Indeed, some of them have insider knowledge. The frightening phenomenon of firefighters actually perpetrating arson and setting fires is a recognised problem in some countries. In the US, for example, there have been cases of firefighter arson in full-time, part-time and wildland fire departments. Perhaps the most intriguingly bizarre of these stories is that of John Orr, a full-time firefighter and former Chief Arson Investigator in the Glendale Fire Department in California. In 1991, a federal Grand Jury indicted Orr on charges that he set or tried to set blazes in eight stores during a three-year period. Four people, including a child, died in one of the blazes. There were countless other fires where he was a suspect. He was convicted but it is probable that Orr would not have been caught except for a fingerprint on a partially burned incendiary device used in one of the arson attacks. There was also the fact of Orr's arrogance. Before the investigation, Orr asked his secretary to type a letter to several book publishers as he had just written a 'novel' about a firefighter who turned into an arsonist. He wanted to pitch the book as one about a real-life serial arsonist, saying, 'he has not been identified or apprehended, and probably will not be in the near future. As in the real case, the arsonist in my novel is a firefighter.' Orr was sentenced to life imprisonment and the jury was deadlocked on whether to recommend that the former investigator should be executed.

There are many other cases, and American research on the issue is extensive in particular. There has not been a case, reported, of firefighter arson in Ireland but one can understand that lure that the fire service would present to an arsonist, particularly one with a deep-rooted fascination with fire.

Where there is arson, there is, of course, arson investigation. Ireland is at an embryonic stage in this area, but the whole discipline of arson investigation and indeed fire investigation in general is set to expand rapidly in this country. March 2005 saw the establishment of the Fire Investigators Association of Ireland, which is now part of the International Association of Arson Investigators (IAAI). At its first session firefighters and fire officers from Dublin, Limerick, Galway and Kilkenny were in attendance.

According to Paul Carolan, the development of fire investigation in this country should see a meaningful difference in terms of resolving the number of undetermined or unknown causes of fire. 'There have been, on average, about 35 per cent of fires in this country where the cause is not known or is not definite. Fire investigation and proper training will make a huge difference to the confidence of fire officers when it comes to making an origin-and-cause decision.' Successful integration of fire investigation into the fire service would bring a quasi-legal element into the work of firefighters, though this would not have any bearing on how they actually do their work on a firecall and would only be used as an investigative tool.

Jamie Novak, a fire investigator from St Paul in Minnesota, told Irish firefighters at a recent conference that since they are often first to the scene, they may witness evidence that could disappear or be destroyed as a fire progresses. 'Sometimes it's easy to prove that a fire is arson, but not to prove who did it. Fire investigation is all about determining the correct origin and cause of a fire.'

While the reasons behind someone's compulsion to deliberately or maliciously set a fire range from the bizarre to the narcissistic, the actual reaction of humans to fires danger is equally baffling. At the third International Symposium on Human Behaviour in Fire, in Belfast in 2004, delegates were told that usually, the general public underestimate the dangers of fire. This finding may be expected, but it is corroborated by people's alarming lack of knowledge of why exactly fire is so dangerous. In an article in *Fire* magazine in 1989 it was revealed that: 'it has been estimated that in 90 to 95 per cent of all cases smoke is the main cause of death rather then heat. Direct flame contact and burns are rare and most victims die in rooms distant from where the fire originated. Often the hazard is not recognised early enough or there are no safe escape routes available. Compounding this problem is insufficient understanding of the real ability of fire detection and suppression systems to quickly reduce the adverse effects of fire. Knowledge of how to use fire safety equipment is often incomplete, even though occupants might have the equipment at home.'

At the symposium in Belfast, those involved in fire prevention might realistically have hoped that times would have changed from this 1989 article, however the presentation by Rob Taylor and Sharon Pepperdine of the Metropolitan Fire Brigade in Australia confirmed that current trends were not much, if any, of an improvement. They quoted 2001 source material from *The Fire Prevention and Fire Engineers Journal* which questioned the whole policy of advocating smoke alarms as the way the alarms are regarded is a good indicator of how seriously people consider the risk of fire. Evidence confirms that many smoke and fire alarms are poorly maintained and a fair quantity are totally redundant, with no batteries. Elsewhere in the home, the average human is blissfully aware of many of fire's causes and dangers, with faulty wiring, having none or poorly maintained fire-suppression equipment and a very low level of confidence in how to use it. In the Australian presentation, in a First World country like Ireland, less then one third of those featured in a survey had some sort of pre-planned idea about what they might do in the case of a fire.

The type of fire which is obviously most likely to be encountered by the general public is a domestic fire. It is a fact that most fires are discovered soon after ignition, but prior to actual flame there may already be smoke. Studies into human fire behaviour indicate that people will often investigate an alarm or what is causing smoke rather then evacuate. It is also a stark fact that human error plays a massive part in fires. The symposium was told that in 96.4 per cent of residential fires investigated, someone was at home at some stage during the fire: 'fires often resulted from seemingly unremarkable events such as cooking or using candles.'

In a study carried out by the University of Ulster, also presented at this symposium, it was revealed that, in many cases, the time taken to start an evacuation is often greater then the actual time of the evacuation. This may have a lot to do with modern lifestyles, where absorbing forms of entertainment have a deleterious effect on people's ability to actually realise that they and those around them are in stark and growing danger. Once the fire becomes recognised, a person will go into one of two modes: to flee or to fight. Of course, this depends of the severity of the fire, but a person's ability to appreciate this may also be dependent on their level of fire-prevention and firefighting knowledge, and of course what fire-suppression tools are available to hand.

In the study presented at the symposium, it was discovered that people will rarely grapple with a fire that they are not confident they can handle. So, in spite of all society's failings in preparing for fires, learning about fires or ensuring that they have equipment for being alerted to or fighting fires, when confronted with the beast itself, people are quite rational. The common notion of total panic when confronted with fire is apparently not always the case. A 1996 article in the *Fire Engineers Journal*, entitled 'People in Fires', said that panic was the exception in human behaviour in fire situations: 'It was found that people tended to behave in a controlled and rational way in fire incidents.' However, as an average of over 50 people die in fires in Ireland every year, people can be controlled and rational and still do the wrong things. The Farrell Grant Sparks report revealed that, based on 1996 - 1998 data, Ireland was at the upper end of the spectrum as regards total fire deaths per 100,000 population.

Fire Behaviour

While the dangers of flame, fire and smoke are all too apparent, the actual nature of fire is hard to define. So what is a flame? A flame is actually the point in which a heat-releasing reaction between fuel vapours and oxygen takes place. When this happens, light is also produced, generally orange or yellow in colour. This is what we recognise as a flame. For this flame to exist, combustion must occur. The fuel must be heated to a point where it produces a vapour which will react with the oxygen and then burn. But with such an unpredictable element, there are many different reactions which will produce many different types of flame.

There are two distinct types of flame, pre-mixed flame and diffusion flame. Both have different properties and characteristics but are everyday occurrences. The key to the burning process and the growth, power and destructive force of flame is that the entire process is self-sustaining. Simply put, a certain amount of heat energy is needed to start a fire. In the fire behaviour training manual used by Dublin Fire Brigade, the process by which a candle is lit is explained. 'When a match is held close to the wick of an unlit candle, the wax melts and rises up the wick by capillary action. There, it evaporates and a flame is established at the interface between the evaporating fuel and the surrounding air. The fuel and air are not mixed before burning, so this is a diffusion flame. Once the flame is established the process of melting, evaporation and burning is self-sustaining.' Diffusion flames exist at the point where fuel and air collide. Unlike pre-mixed flames, flammable vapours do not exist prior to burning. The dominant process here is the mixing of the fire and oxygen molecules, triggered by heat. A candle is an example of a slow-burning diffusion flame.

Pre-mixed flame is far more complex in how it comes into existence. Technically speaking, pre-mixed flames occur when a gaseous fuel is well mixed with an oxidant, normally air. For ignition to occur a flame or spark must be applied to this mix. Of course, there is plenty more science in play when it comes to explaining exactly how this fire will burn. Suffice to say not every mixture of air and fuel will actually burn; the mixture must be within the flammability limits which are different depending on the gas which is mixing with the oxygen. Some gases, like hydrogen, have a broad flammability limit while others such as methane have a very limited flammability range.

The temperature when such a mixture ignites is known as the flashpoint. For instance, take a pool of flammable liquid such as petrol. The petrol emits flammable vapours which mix with the air. When and if the temperature passes a certain critical point, known as the flashpoint, the introduction of a spark or flame will ignite the liquid. This is why at petrol stations you will often see signs asking you not to use mobile phones at the pumps, because mobiles are not 'intrinsically safe'; this means that they may produce a tiny ignition point which could ignite flammable vapours. Other common devices that could provide this spark include refrigerators and televisions.

Fire and the nature of flame is notoriously difficult to predict. Modern firefighting is a mixture of craftsmanship and science. The trinity of elements which contribute to any fire — fuel, air and heat — can combine in a multitude of ways, creating a range of fire phenomena to challenge the modern firefighter. The advent of training in these disciplines was provoked by developments in Swedish firefighting following the death of several of their personnel in the Eighties. What followed has developed into standard training for any modern brigade, combining tactical know-how with scientific savvy.

According to Paul Carolan, Health and Safety Officer with Dublin Fire Brigade, while the number of fires to which fire brigades are called out has consistently decreased, the threat which these fires pose to firefighters in terms of heat, intensity and power has increased dramatically. 'Today's structures are more resistant to fire then ever before but, in contrast, when they burn, they do so with a far greater degree of thermal-radiation intensity then before. Insulation and man-made fibres contribute to changes in fire behaviour and in the nature of fumes which those fires give off. Today's fire is hotter and more toxic then ever before.'

While the existence of backdraught, flashovers, fire-gas explosions and other fire phenomena are accepted parts of firefighting reality, learning how to deal with them is a relatively recent development. When the deadly mechanics of these phenomena become clear it is a wonder that more firefighters were not lost due to a lack of knowledge of the dangers in previous decades.

In fact it took a lethal event to drive forward change in how complex fire phenomena were viewed by firefighters both here in Ireland and elsewhere, that lethal event took place in the village of Blaina in Wales on 1 February 1996 and led to the death of two firefighters in a backdraught situation. The event is still used as the case study for training in this area today.

In the official report into the tragedy at Blaina, the devastation of the explosive force and the extreme temperatures of the backdraught at 14 Zephaniah Way is depicted in photographs of the firefighters' equipment. One fire helmet is reduced to a charred lump and the tunics of the firefighters who perished are little more then shards of blackened material. The lives of Steven Griffin and Kevin Lane were lost in a complicated delayed backdraught in a conventional residential fire. Flammable gases from a fire which originated in the kitchen of the house escaped into surrounding areas of the dwelling by burning through the ceiling of the kitchen. The seat of the fire was in the hot press. The backdraught ignited approximately four minutes after the two firefighters had initially entered the building, searching for people on the upper floor. They successfully rescued a child but were informed, mistakenly, by a member of the public that a second child was still inside. The fact that the fire was in the kitchen had led to the firefighters' mistaken perception that they would be relatively safe elsewhere in the house because it was smoke logged. The movement of Griffin and Lane through the house introduced oxygen into an area where hot fire gases had escaped from the kitchen.

The firefighters who perished in the blaze at Blaina had no fire-behaviour training and were unable to recognise flashover or backdraught symptoms. Blaina had a defining effect on fire-behaviour training yet it was still almost four years after this terrible event that such training was pioneered in Ireland.

Eddie Reilly, who is heavily involved in training recruit firefighters to deal with these events at Dublin Fire Brigade's training school, the O'Brien Institute, explains that the fire-behaviour aspects of training are vital both to provide proper knowledge of potentially hazardous situations and to dispel the myths surrounding the unknown. 'While it was accepted, prior to fire-behaviour training becoming commonplace, that things such as flashover and backdraught existed, there used to be six or seven definitions to differentiate between the various symptoms, but you could put them into three categories now: backdraught, flashover and fire gas explosion.'

The most commonly known of these, perhaps due to the movie of the same name, is backdraught. It is recognised as one of the most terrifying and potentially deadly events which a firefighter can encounter. 'A backdraught is extremely powerful and potentially deadly,' explains Reilly. 'In a situation where a backdraught develops, the fire has been starved of one of its three essential elements, in this case, oxygen. The fire will generally originate in a contained area where at a certain point all oxygen in the room is consumed and the fire begins to decay, literally smothering itself.

'There is no oxygen to help ignite the fire gases so they develop exponentially, literally filling the compartment. The fire is at this stage dormant. What it needs to re-ignite is the presence of oxygen, to mix with the fire gases and create what is often referred to as the "ideal" mix.' For any firefighter entering a structure where this scenario may be developing, the paramount need is to avoid introducing oxygen into this lethal mix. If oxygen does enter the fireground, sooner or later, this 'ideal' mix of heat, gas and air will occur. 'It can happen suddenly or, like in Blaina, it can take an indeterminate amount of time, but it will happen. There will be an explosion, and the fire will rush forward from ground to ceiling, engulfing anything in its path. It will also have tremendous kinetic power in the form of a pressure wave as it blasts through the opening from where the oxygen came.'

Thus, backdraught is a fire in decay due to lack of oxygen and this decaying fire can become potentially explosive due to a sudden increase in the presence of oxygen. This can be caused by a door or window

breach or, as occurred in Blaina, compartment failure. Recognised signs which indicate the possibility of backdraught are the following:
— smoke-blackened windows
— smoke billowing from the eaves of a building
— heavy presence of thick, dark smoke
— smoke pulsing around doorways or other entrances.

Less violent but still potentially lethal is flashover, as Eddie Reilly explains. 'In flashover a fire is able to reach full development due to the heavy presence of the three principal elements; fuel, heat and air. In a situation where a flashover occurs there will be an increased rate of gas emission from super-heated surfaces due to the high temperatures. Surfaces are searingly hot and a smoke layer will form in the room, essentially unburnt fire gases. Ceiling-level flames also occur in a flashover as they burn through the gases which rise.'

Today, the training for firefighters in dealing with flashover, backdraught and other phenomena such as fire gas explosions is thorough and intense. 'One of the first questions we often ask our firefighters is what they would do if they walked into a room full of petrol fumes,' says Eddie Reilly. 'The correct answer is, of course, to ensure that you take the precautions which you are trained in and that you have proper fire cover. Before fire behaviour training it was a very common situation for firefighters to walk into a room and see smoke but presume that there was no danger because there were no flames present.'

The DFB fire-behaviour training takes place in a modified container structure, similar in size to a shipping container but with unique and subtle additions. Firefighters must complete three exercises in the fire-behaviour course. In the first exercise, firefighters are taught about fire behaviour, witnessing a fire developing from the start to full development. The students are brought into the container in batches of eight, in full BA. A fire is lit and the students observe its development. During the demonstration, a layer of fire gases will form in the ceiling and then fill the container down to approximately three to four feet from the floor. The students are taught about the negative pressure which now sucks in air from outside and the positive pressure which expels gases from the container. The gases then begin to ignite and students observe how different gases ignite at different temperatures.

As the gases start to ignite, the students are withdrawn from the container and witness the flashover from outside. Instructors then demonstrate how to control the fire using water-pulsing techniques. The second exercise is the flashover attack. Instructors bring the fire to full development before the students enter. They are brought into the container in teams of two and use the pulsing techniques which they have been trained in to control and extinguish the fire. The third exercise is a closed-door exercise known as backdraught attack. In this exercise, instructors light a fire and close down the container to create conditions which are conducive to backdraught. Firefighters then effect door entry procedures which they have learned to enter the container safely and proceed using their trained techniques.

In today's firefighting environment, health and safety are paramount. The situations in which firefighters are placed during training are not unpredictable, as, due to the high instructor-to-student ratio, they can be monitored closely, but they must constantly be aware of the threat of accidents. On the fireground, absolutely every square inch of flesh must be covered. This is in contrast to the older, unofficial, code of firefighting which advocated leaving the ears exposed so you would know when it got too hot and it was time to retreat. 'It's very much safety first now, which is only right. No skin should be exposed at all on today's fireground and nobody will enter the training container without a fire hood,' explains Eddie. This is doctrine borne by bitter experience. In 1999, when firefighters tried in vain to rescue a child at a fire in North Dublin, only one of the crew of nine had his head covered properly. The other eight men had to go to hospital for treatment to nasty facial burns.

The essential reason for fire-behaviour training in any fire brigade is to learn something of the mechanics of fire development. Recruits and students are taught the chemistry of combustion and how various fire phenomena develop. They are taught how to deal with flashover and backdraught in addition to the dangers of potential gas explosions in adjoining rooms. 'This danger is not always recognised,' says Eddie Reilly. 'If there is a fire gas leakage into a room next to a fire all it takes is for someone to hit a switch which ignites a spark and then the whole place can go up.'

'Variants of flashover and backdraught are things I've encountered many times,' says former firefighter and fire consultant Frank Cafferty. 'It wants air and if you're a firefighter, you will, at some stage, be standing at a locked or closed door, wondering what exactly waits inside. What we used to do was use two of our most basic functions, touch and listen. If the handle of the door is hot, for example, you know there is a fire inside. Also, you can hear the high-pitched noise of the air being sucked in towards the flame, through the door.'

Cafferty, now retired, worked with Roscommon Fire Service for almost thirty years and prior to that he served with Leicester Fire Brigade in England. He remembers his first experience with fire phenomena, at a factory fire in Leicester city in 1962. 'When we arrived at the fire, firefighters had already entered the building. The fire was at night, a Sunday, I remember. The factory was a four-storey structure which manufactured thread and yarn. That meant plenty of fuel for any fire, but the problem on this fireground was not flame but smoke. The guys inside couldn't find the fire to fight. We had several firefighters inside complaining of smoke-inhalation so it was then that the fire officer on scene decided to pull the men out. I remember being on a roof to the rear of the building with a hose line when suddenly, the ground floor exploded into flame, with a roaring sound. The first floor, the second, then the third and the fourth. The whole building was suddenly engulfed in sheets of flame.' The firefighters who had only twenty minutes previously been in the building were extremely fortunate. 'What they had in fact been walking through in that building were clouds of fire gases waiting for an influx of air to ignite the blaze. The seat of the fire was in the basement to the best of our knowledge. Once air became available, the whole thing erupted again. What happened that night is something which we refer to as a "mushroom" flashover.'

With the high-spec personal protective equipment of today though, some of the more basic senses — such as touch and hearing — are compromised by safety standards. It's an understandable trade-off and firefighters do have access to devices such as thermal-imaging cameras which allow them to see within rooms and buildings and find the heat sources.

However, for firefighters, even if they are better equipped then ever to deal with this most hostile of environments, it can represent an instant confrontation with mortal danger. On 21 November 1979, two DFB firefighters, Michael Loughlin (RIP) and Noel Kelly came face to face with the terrifying power of a flashover. A unit from Dolphin's Barn station was called out to deal with a fire at a plastics factory in a west Dublin industrial estate. When firefighters arrived on scene, the blaze was, in firefighters' terminology, 'well developed', with dense smoke pouring from the building, leading to the evacuation of offices and factories around the fireground. There were two appliances on scene, one was at the rear of the factory and had begun to apply two hose lines to the fire. The other, with Kelly and Loughlin, was at the main entrance. 'The D/O told us to go inside and check the premises for people and to vent the building, as was procedure at the time,' remembered Michael Loughlin when interviewed about the incident for *Firecall* magazine. 'By opening the shutter at the front of the structure we thought that we could vent most of the smoke and we had the initial impression that it would be a fairly straightforward task.' Fire safety procedures have radically changed in DFB since this incident, where the dangers were seriously underestimated. The firefighters entered the building wearing BA and made their way in dense smoke through an office and then turned left into the main floor of the factory where masses of boxes and crates were stacked. Both men were conducting a search for anyone who may have been overwhelmed by the toxic smoke when a roar and a huge flash rose up from the back of the building. The flames roared down the walls and ceiling towards Loughlin and Kelly. Michael Loughlin remembers their frightening speed. 'Before we actually understood what happened the flames were raging around us as we cowered on the floor. Literally, we were frying. My hands felt like they were melting, like someone had submerged them in acid. The pain was unreal. I could feel the hairs on the back of my neck singeing and burning, we were both terrified.' Michael also remembers feeling a strange urge, amidst the suffocating heat, to remove his BA mask. 'It was obviously completely irrational and would have been suicidal but I remember just giving myself a second to calm down and get a grip on myself.'

After the initial shock, both men looked around for an escape route, but everything was ablaze and they were surrounded by flame, smoke and searing heat. They were completely trapped. 'It was a desperate situation. I remember then, through the pall, seeing a gate with about an inch-and-a-quarter gap at the bottom. It was like a thread from which our lives were hanging.' Loughlin thrust his hand through the gate, screaming for first aid in a voice that could not possibly be heard on the outside. 'I was hoping against hope that someone would pass through a first aid package to help cool us down

from the flames. Next thing, a sledgehammer was passed through!'
Fumbling in the chaos, Loughlin set his hand upon a set of keys to the
gate but they also turned out to be useless as there was a roller shutter
blocking the way.

As both men tried to remain calm beneath the fire which raged
above them another team of firefighters, alerted by the men's personal
Distress Signal Units (DSUS), were making frantic efforts to get to the
men but were beaten back again and again by the intense power of the
flames. At this stage, desperation began to take hold both for Kelly
and Loughlin and their colleagues outside who were watching
helplessly as the flame consumed the building. Inside, the two trapped
firefighters decided to stay near the gate, where there was a sliver of air
coming through the tiny gap at the bottom. 'We had completely run
out of options. We had to wait either for rescue or for the inevitable,
should the air in our BA tanks run out. It was around this time, when
we had exhausted all our options, that I heard a rumble from outside
followed by a heavy thumping sound. Something was trying to get
through.'

As time ran out for the men inside, the remaining firefighters
outside had been trying to formulate a rescue plan, however
desperate. One of Noel and Michael's colleagues, Mick Smith, dashed
to a nearby building site to ask for help from the driver of a JCB. It was
an unorthodox, improvised masterstroke. 'I remember hearing a
crash and next thing the gate fractured in the middle and created a
small opening at the bottom. Not much, but enough for us to crawl
out through. We were grabbed and pulled through the gap by the
other lads on the crew.' Both men were suffering from minor burns
and shock, but were discharged following brief hospital treatment.

George Harrison, still serving, was on scene at the time. 'We haven't
lost firefighters in a fire in this Brigade since 1936, but it was a close
call that day. Those men were very fortunate. I remember I was at the
rear of the fire when someone came running to me, saying that a DSU
had gone off. We knew exactly where the lads were, but the roller
shutter on the door was operated by a crank on the inside and we
couldn't get to it. Dublin Fire Brigade had recently upgraded our
firefighting kit and I have to say that without that, we could be talking
about a tragedy rather then a close rescue.'

Part Two
Part-time Heroes

The Retained Brigade – Carlow

For their counterparts in the vast majority of Irish counties, station life in the fire service is very different to that in the capital. The retained brigades, which operate at some level in every county, form the vast bulwark of the fire service. The first thing you notice about retained stations is that they are definitely cleaner! Full-time stations never get a minute, one watch goes out and another watch comes in. The stations are maintained and cleaned as part of the job in between fires and other emergencies and they bear the scars. Retained stations are by their nature far quieter places, but with recent investment many of the new stations in rural counties are highly impressive. In Carlow for example, which had a new county fire service HQ opened in 2003, the building, the facilities and the equipment used would compare well with any service worldwide. The station only has one full-time trained firefighter, Station Officer John Comerford. 'Fire Brigades will always be busy, it's all relative to your brigade; we wouldn't have the firefighters we have if we didn't need them.' Carlow, on average, receives approximately 250 calls a year. It's one of Ireland's smaller counties but a growing economic and population centre. There are four stations in the county, Hacketstown, Tullow and Bagenalstown in addition to the main station in Carlow Town itself.

For retained brigades, the work they do with the fire service is a mixture of professionalism and pride. These men — in Carlow there are no female firefighters as of yet — simply love the fire service. 'I suppose it is a different mindset required for a brigade like this. I think that any firefighter, full-time or retained, would only join a fire brigade if they really have a love and passion for the work that firefighters do, but for full-time members there are also economic concerns. I don't think that exists to the same degree in the retained brigades.'

For the firefighters in Carlow town, it is evident that this passion burns strongly. 'It would kill me if I ever had to leave the fire brigade, it's not the sort of job you would ever take up if you just needed extra money, there are far easier ways of making more money, but we do it because we really love doing it and providing the service,' says Paul Curran. The fourteen firefighters in Carlow are a sociable, talkative mix but at the end of the day they are fire and rescue professionals and due to the nature of their brigade, this requires swift reactions and a quick change of mindset. 'When your pager goes, everything stops and you're out the door,' says Pat Bolger, 22 years with the brigade. Retained firefighters all live locally to their station so while geography isn't an issue, the unpredictability of the work can be. The vast majority of retained firefighters have full-time jobs, but once the fire service pager goes, they respond immediately. It can and does place a burden on social life and work life, because when you're on call, you are literally never off duty. 'When the pager goes, first you get the adrenaline rush, the buzz of getting knocked out of your normal routine and knowing that you will be facing a unique situation, since every callout is completely different,' adds Bolger. 'It's like an automatic switch in your head,' says colleague Pat Craddock, 'you get used to it, but your family, your children and your employer have to get used to it as well.' The consensus is that the enthusiasm levels may seem higher in the retained brigades due to the fact that for them, their duties are something out of the ordinary.

While the passion of turning out for the fire service burns brightly all retained firefighters, just like their fulltime counterparts, are exposed to harrowing, distressing and at times dangerous incidents. In the retained brigade, firefighters obviously have less experience in dealing with these incidents, and the scar tissue which protects the emotions of the full-time firefighter so well is not as strong here. In addition, the retained service provides fire and rescue within their own communities so there is an awareness that whoever they are being called out to help may indeed be someone they know.

Paul Curran was called out to a water-rescue incident in which tragically it was one of his own cousins who drowned. 'He was only a two-year-old boy; at the time I didn't know who exactly it was until we got down to the river. His mother had gone in after him and we rescued her, but the boy was dead when we retrieved him.' Dealing with death and tragedy on your own doorstep is not easy: 'It was obviously an awful experience but I think what I have learned from being in the brigade did help me cope and it also helped me help others to cope.'

Like all firefighting and rescue personnel, it is incidents with children which really hurt. 'They're without a doubt the worst, absolutely; any firefighter will tell you about incidents which involved children because they're impossible to forget.' The fraternal support structure, common in the full-time brigades, also exists in the retained but the connection is obviously looser. 'When we're finished a call, we're effectively off duty,' explains one of Carlow Fire Service's more recent additions, Darrell Hayden. 'If we have just gone through something nasty, a very bad RTA for example, I always find it helps if you get another callout again straight afterwards. It just helps you move on, to get what happened out of your head.' Like all fire services, there is now a counselling procedure for retained brigades if and when it's required. However, the crew is not too eager to talk about their own problems after incidents. 'We're trained to do this job and we know we're going to see some things which we wish we hadn't,' says Pat Craddock. 'However, whatever we're going through is absolutely nothing compared to the loss and the pain which some family is going through as a result of a fire or some other tragedy. We're just coming back to the station and getting on with our lives. Elsewhere, some mother is getting a visit from the gardaí with news of her dead son and the grief for them is just beginning. You need to keep your emotions under control or else you're of no use to anyone.'

The retained brigades work because of the dedication of their personnel to their own communities. In the words of John Comerford, the firefighters couldn't be paid for the amount of their own time they put into helping the brigade. 'People don't think about the fire service until they need them and then we're never there quick enough,' says Curran. 'That's true for us the same as it is for brigades all over. People don't really understand how the retained service works, or the amount of work that we are trained to do.'

The retained service is eager to get involved in SRT as there are a number of water-related incidents every year but not the training available to allow the firefighters deal with them properly. In 1982, William Kavanagh, a firefighter with the service lost his life while trying to save another in a water-related callout.

Of course, firefighters could be completely trained from A to Z in every aspect of fire and rescue but still not be ready for everything. Mick Gahan remembers getting a call from the gardaí to help search for an elderly man who had gone missing one cold winter's evening. It was known that the man was suffering from Alzheimer's disease. 'Things like that are obviously not in the training manual,' says Gahan. 'We are trained as medical first responders but not in how to deal with mental disorders. Nevertheless, we do our best and we were looking for this man for hours.' The firefighters finally found him, shivering under a bush in a park after hours of searching. He recovered, but so low were the temperatures that he wouldn't have lasted until morning.

Between fires, rescue and search and rescue there are also the callouts which reflect a sad shadow of society, which Willie Bermingham shone such an effective light upon during his lifetime. For some time, the Carlow service had been getting calls from an elderly lady in the area. The gardaí had referred her on to the fire brigade and the CAMP controllers in Dublin had put a call through to s/o Comerford to see if he could help: 'Basically she's just lonely; we've got several calls from her at this stage.' Indeed the firefighters had been at her house before, where she said someone was trying to gain entrance. In the words of the crew she was just lonely and wanted reassurance, so they diligently checked around the house and even changed some light bulbs and plug sockets. 'We were going through the motions for her, we knew there was nobody in her house but it's very sad to see an old lady like that,' says Pat Craddock, 'you do see some very sad aspects of society in this job.'

Edgeworthstown

On Friday, 21 April 1995, around 9 pm, a Scania truck carrying 20,000 litres of liquefied petroleum gas (LPG) was travelling near Edgeworthstown, Co. Longford when the driver braked suddenly, sending the vehicle out of control, skidding along a bend, smashing into an electricity pylon and then crashing through the front garden of Mrs Theresa Hughes. Tragically the driver, Dan McCartan, was thrown clear from the cab of the truck and died soon after, but the scenario facing Longford Fire Service and other emergency crews due to the cargo which the truck was carrying was the start of a 24-hour nightmare.

As the truck ploughed through the wall of Mrs Hughes' house, it careened onto its side and, according to witnesses, the air around the vehicle began to fill with gas immediately, spreading rapidly in the still evening air. Apparently, a valve at the top of the truck's gas tank had become damaged during the course of the accident and LPG was leaking freely. The first to raise the alarm were children who were playing football nearby and when gardaí and firefighters arrived, after tending to Mr McCartan, Chief Fire Officer Vincent Mulhern and Station Officer Michael Smith initiated the Major Emergency Plan. Essentially, a Major Emergency Plan is a county-wide formula for responding to significantly threatening man-made or natural events. It involves civil authorities, law enforcement and emergency personnel and each part must be implemented properly in order to elicit the desired overall response.

There was a serious leak of explosive gases and all persons within seven hundred yards of the leak had to be evacuated immediately. These included the sixty-one residents of a nearby retirement home who were rushed to nearby hotels, which acted as temporary refuges. The entire Longford Fire Service was put on alert for the incident, with over forty firefighters on scene. Just over an hour later, as firefighters struggled to contain the scene, the gas which had escaped from the truck and seeped into the Hughes household ignited in a terrible explosion.

Pat McLoughlin, a local man, was only about thirty yards from the area when the explosion occurred: 'When I heard about the crash I went down to see what was going on. There was no fire at the time, but when I was about thirty or forty yards from the scene it just blew up. It was more a thud than a bang and it blew me back a couple of yards.' The firefighters now had to deal with a fire in addition to a gas leak. The house of Mrs Hughes, who had recently been widowed, looked like it had been subjected to a flamethrower attack.

Throughout the night, as flame gushed from the severed valve on the gas tank, fire crews frantically but successfully fought to keep the tank cool. At the time there were rumours that the tank itself had exploded. If this was the case, a significant part of the town, in addition to the majority of Longford's firefighters would have been wiped out. Until the fire at Albert Reynold's C&D Pet Food plant in January 2006, this was one probably one of the most serious incidents faced by Longford Fire Service. At its height, 250 people were evacuated from the area and stationed in hotels and other public places at a safe distance. Initially, reports the *Longford Leader*, there was scepticism as to the severity of the incident and that the implementation of the Emergency Plan had been a gross over-reaction. The fact was, however, that this was an extremely serious incident. The unfortunate death of Mr McCartan and the destruction of the Hughes holding were tragic, but minor compared to the price which Edgeworthstown could have paid if the scenario had played through to its worst-case conclusion.

Cow Comforts

The massive fire which engulfed the Cow Comforts factory in Belmullet, County Mayo, on 1 April 1999 caught an entire community unawares, catapulting firefighters into a prolonged battle with toxic flames which lasted over twelve hours. Padraig Conroy, a firefighter on the scene, told the *Irish Times* of Saturday 3 April: 'It was like going into a chamber of hell. The place was literally engulfed in flames, spreading like a volcano of fire. We were fighting a fierce fire until 4 am on 2 April.'

Belmullet had seemed quite a different place just fourteen hours before. Chris Ruddy, s/o with Belmullet Fire Brigade (now retired) was winding down to the Easter break on 1 April 1999. It was Holy Thursday, the prospect of a long weekend ahead. It was also April Fools' Day, when everyone was prepared for at least one practical joke. Chris had served with Belmullet Fire Brigade for over twenty years, and his work with the County Council kept him practically next door to the town's fire station. The station itself, opened in 1986, is a modern two-bay structure with good facilities and a decently sized drill yard and training tower. He presided over a well-equipped rural retained brigade, himself and nine other lads at the time, including Sub Officer Joe Murphy. On average, he would expect about a hundred callouts per year with road traffic accidents, house fires and gorse fires the traditional culprits. That afternoon, as lunchtime slipped into early afternoon, nobody in Belmullet, a town of approximately a thousand people, could have foreseen what a bizarre, frightening and potentially tragic night lay ahead.

Near the town's fire station, a few hundred metres up the road, a piece of pipe carrying hot oil to the pressing area of the Cow Comforts rubber mat factory finally ruptured and burst, spraying scalding hot oil into the pressing machine. Here, the oil ignited and triggered a blaze that would destroy the factory, lead to the entire town being evacuated and would severely test the resources of Mayo's fire service.

Cow Comforts was located in Belmullet Industrial Estate, about half a mile from the centre of the town. This factory had a manufacturing area of approximately 2,969 square metres and on the eastern side there was a two-storey administration area which included canteen and toilets. Prior to being operated by Cow Comforts, this factory was involved in rubber shoe-sole production, operated by a Spanish company named 'Solano'. This company ceased trading in 1992 but its production methods were similar to those used by the next firm to operate the factory, Cow Comforts. The brainchild of a former Solano employee, Tom Duffy, the new firm would be involved in the production of rubber mats for cattle. These mats were manufactured for the domestic and export market and the firm enjoyed a successful launch, growing on an annual basis until stopped in its tracks by the terribly destructive fire of 1999.

The machinery in the factory was quite similar to that used by 'Solano', essentially, the new firm produced a variant rubber product which meant good news for Cow Comforts but bad news indeed for County Mayo Fire Service. The factory itself was, in the words of Chris Ruddy, 'an accident waiting to happen'. Sustained years of rubber melting and moulding for both shoe soles and now rubber cow mats meant that the interior of this structure had been coated and recoated with a film of hardened rubber soot. Easy to burn and toxic to boot. There was also the sheer amount of mats that were stocked there. The factory used to operate in a certain way. In the autumn and winter the mats would be produced and then in the late spring production would cease and the place would be used as a warehouse for the company to offload the stock from. This meant that the workforce was made redundant during the summer. When the fire happened, the factory was nearing the end of the production cycle and the place was well stocked with these synthetic rubber mats.

During the production process, the mats were rolled from a rubber dough and cut to a particular size. They were then stacked in batches of three or five and placed in a mould which was then placed in a press. In the press, this material was subjected to extreme pressure equivalent to 1,300 tonnes and was then super-heated to 160 degrees centigrade. It was here, at this point in the process, that the fire ignited. 'To be honest, when it happened we weren't surprised there was a fire, just the scale of it,' says SUB/OFF Joe Murphy. 'It wasn't the first time we had to attend a callout there and the problem which ignited this fire had been spotted before, during a previous callout, when a burst pipe led to oil igniting which in turn led to a boiler exploding which blew a segment of concrete out of the building.'

That spring afternoon, when the pipe burst and the superheated oil gushed into the pressing machine, human tragedy was a distinct possibility. This was an angry, violent start to the fire and Chris Ruddy got quite a shock with his first sight of the inferno. 'When I walked out of the station after receiving the alarm, I remember seeing a huge, black cloud of billowing smoke rising high into the clear sky.' Under

Chris's instruction, Belmullet's firefighters prepared to respond to the blaze but in the station officer's own words, 'We're a modest two-engine station, with ten firefighters including myself and the sub officer, we knew we weren't going to be able to handle it ourselves.'

Almost an hour's drive away, in the county capital of Castlebar, Mayo CFO Seamus Murphy received a call at 14.38, just six minutes after the initial alarm had been raised. He mobilised his brigades in Castlebar, Crossmolina and elsewhere in the county and, along with his senior officers, left for Belmullet immediately. At 14.36, Belmullet Fire Brigade was on the scene of the inferno with MO 21 A1, the callsign of one of the station's engines. The smoke was thick, black and toxic so BA was a must for all firefighters. Two men were sent to the water pump house to ensure that there was the optimum pressure required for the battle ahead, meanwhile the remainder of Belmullet's ten-man crew started fighting the fire. 'We were in the thick of it fairly quickly,' recalls Joe Murphy. 'We learned as we arrived on scene that there was nobody in the building and once that was verified it removed that particular worry, which is always paramount when you arrive at a fire.'

Chris Ruddy remembers that while that was the good news, there was plenty of bad news. 'The flames had already penetrated the roof, orange flames shooting into the black smoke and it looked like the roof might collapse at any time. We were dealing with a possible structural collapse and a toxic blaze, so we knew straight away that sending firefighters into the building was not an option.' Belmullet's firefighters started hitting the fire with everything they had from both the southern and eastern sides of the factory.

Meanwhile, as Belmullet darkened beneath the rapidly advancing smoke plumes, firefighters from around the county were preparing to assist their colleagues. As CFO Seamus Murphy had set out towards the fireground, stations in Ballina, Crossmolina and Castlebar were despatching their own vehicles as backup. Amongst these was a chemical incident and communications vehicle from Castlebar. This is a specialist emergency tender, not possessed by smaller stations like Belmullet. With toxic smoke on scene and the Mayo Major Emergency Plan likely to be activated, it would be needed at the fire. Ballina sent a hydraulic platform, known as a 'Snorkel', which would provide a much-needed height advantage and help firefighters hit the 'seat' of the blaze at Cow Comforts. When CFO Seamus Murphy arrived on scene just after 16.00, four firefighting tenders, the Chemical Incident Vehicle, the Hydraulic Platform and a 4x4 transport were on scene. The initial plan was to try and slow the progress of the flames where they were most intense, on the southern and eastern sides, and a Breathing Apparatus control station was located here.

The plume of black toxic smoke was at this stage menacing the southern edge of the town and gardaí were advising residents within the path of the smoke to either stay indoors with doors and windows closed or leave the area. By this stage, Mid West Radio had been requested by fire brigade control to broadcast a message to residents of Belmullet to keep their windows closed. It was a harbinger of things to come.

On the front line of the firefight, the situation was clearly getting worse. The billowing black smoke and intense temperatures meant that firefighters were making very little headway in bringing the fire under control. As a structure, the Cow Comforts factory was destroyed and so senior officers on the ground began to discuss alternatives to putting out the fire in a conventional manner. At this stage there were firefighters from five stations around the county in attendance at the scene.

Joe Murphy remembers that in addition to the major challenge of knocking down a fire of this magnitude, the crews also had ancillary dangers to contend with. 'On the southern side of the building, where the fire was at its peak, there were two large diesel cylinders situated quite close to the wall. They were a major concern. If the wind changed and flames reached them, there could have been a nasty explosion. Therefore, we had to protect these tanks, keeping them cool at all times. To do that properly we needed to dedicate personnel to the job, draining our resources from trying to bring the blaze under control. We also had a man, at times two men, monitoring the water pressure in the pump-house. At times we couldn't see them due to the black smoke sweeping over the small building. We broke strict BA procedure there, but we needed men in that spot, I still wonder to this day what could have happened to those men if events on the fireground had taken a turn for the worse.'

CFO Seamus Murphy remembers that as late afternoon gathered towards evening, he and his senior officers would have to take some drastic action. 'By 17.00 it had become clear that despite our best efforts, the progress made in extinguishing the fire was very slow. The southern wall of the building was becoming increasingly unstable which made it even more difficult for firefighters to get close to.'

The blaze was now really beginning to flex its muscles. Chris Ruddy remembers hearing terrible rumbles and loud bangs as the acetylene tanks within the building finally exploded. 'I remember, in addition to the explosions inside, hearing the building itself begin crumple under the heat.' Metal girders finally began to succumb to the flames, wilting and twisting under the immense heat. Seamus Murphy took stock of an increasingly desperate situation and together with the ACFO Austin Gannon and Belmullet S/o Chris Ruddy, the decision was taken to requisition a heavy digger to assist the firefighters by essentially knocking down the stricken building.

'It wasn't a decision taken lightly, I remember that,' says Seamus Murphy. 'We knew from taking stock of the situation there that we couldn't save the factory and this was a choice we made in order to safeguard our firecrews and more importantly, get to a point where we could actually extinguish this fire.' For almost three hours, firefighters had hit the blaze with a ferocious amount of high-pressure water jets. All in all, over the course of the firefight, 1.5 million gallons of water were used, 635,000 in the first nine hours. Despite their best efforts, the roof of the factory had begun to collapse, the southern side of the factory had become almost inaccessible and it was widely accepted that the decision to demolish the factory was the correct one. At 17.06, the call was put out for local access to a heavy digger.

Other events were also beginning to conspire against emergency crews. The pall of black, toxic smoke which for most of the afternoon had been carried by the wind away from the town centre was now beginning to turn. A westerly wind brought it sweeping in over Belmullet town, turning early evening into a premature, eerie dusk. An impromptu summit was held between senior fire officers, local authority officers and senior gardaí. The debate centred on whether the Major Emergency Plan for the county should be implemented. 'When things like the Major Emergency Plan are drafted, you know they won't be used frequently but there is absolutely no margin for error as they may have to be put into action in a highly-charged, chaotic atmosphere such as the one that day,' Seamus Murphy recalls.

On that day, at 18.15, the CFO carried out this action, meaning that the town of Belmullet would be evacuated as a precautionary measure. Back at the fireground, five fire engines and thirty-five fire brigade personnel were in attendance, with the fire still burning out of control. Just after 19.00 a reappraisal of the firefighting plan took place, with one major modification as a result. The operator of the construction machine which had been requisitioned to demolish the building was told to start knocking the severely damaged southern wall of the factory. 'We had two hoses covering the cab of the vehicle at all times to keep him safe. There were also people worrying about him too,' recalls Joe Murphy. Three jets were positioned to protect the machine as it went about its task, eventually tearing down the entire external wall of the production area. It was still proving extremely difficult for firecrews to extinguish the flames though, with the fire burning through the centre of the bales of rubber mats and reigniting spontaneously when it was thought that one was extinguished. Seamus Murphy remembers that the only process that worked was a slow and deliberate one. 'We would get the digger to lift the burning mats and then train jets of water onto it, soaking it from each side. Using this method we made slow deliberate progress until midnight.'

While the firecrews laboured within the now virtually destroyed structure, with the blaze still raging, in the town itself a mass evacuation was underway. A loudspeaker on a garda car was used to deliver the message to an increasingly frightened population. The community was now dealing with both a dangerous fire and a civic emergency. Most people left as they were told, although there were some cases where residents chose to stay. The authorities were powerless to force people to leave. Two evacuation centres were set up, in the school outside town and in the Palm Grove nightclub, as it was known then. Local Ambulance control also needed to evacuate the local hospital; indeed the EMTs actually delivered a baby during the course of the evacuation.

By 10 pm, the town of Belmullet was in complete darkness, swathed in black smoke. On the fireground, the battle for control continued but the efforts of the fire brigade were beginning to pay dividends. There were now seven fire engines and fifty firefighters assisting in the effort. The water jets were moved to counteract the change in wind direction and the mechanical digger also changed its operational attack position during this period. By 2am on the morning of Good Friday, 5 April the tide had turned and the majority of the blaze had been extinguished, with sporadic reignition. The decision was made at this stage to scale down the size of the firefighting operations and the deployment was reduced to three engines with twenty personnel remaining on scene throughout the night. Throughout that long night, fire crews persistently attacked the flames at their by-now-infamous southern flank and with the aid of the mechanical digger consolidated their progress. Meanwhile another crew hit the fire from its north-eastern side. By 7 am, the battle was all but over, the fire was hemmed into a small area of the north-western section of the factory. At 7.30 that morning the Major Emergency Plan was stood down after consultation with gardaí and the confused, jaded residents of Belmullet were permitted to return to their homes.

The gardaí had to conduct a forensic examination of the scene after the fire had been knocked down. 'We knew, already, what had most likely caused this fire and sure enough the forensics team found the piece of punctured pipe which had sprayed hot oil into the pressing machines,' said CFO Murphy. 'That's what started the whole thing. They still have it in the forensic science laboratory in the Phoenix Park, indeed it was the subject of a lecture at a recent conference. That little piece of pipe caused us a lot of trouble.'

Kilkenny Town Hall

'I remember the fire as being particularly spectacular but thankfully it never got particularly dangerous,' says Kilkenny s/o Joe Traynor. The images were certainly striking, flames shooting through the roof and windows of Kilkenny City's iconic town hall. Also known as a Tholsel, the building is recognised as one of the finest of its kind in the country. It was constructed in 1761 and in addition to its distinctive clock tower it also has a unique arcade on the front of the building. Inside, there was a Georgian council chamber on the first floor and the offices of the council themselves.

When the flames were first noticed on the upper floor of the city hall, by Town Sergeant Joe Stapleton, his alertness saved a particularly precious part of Kilkenny's history as he dashed upstairs to the council chambers and retrieved the charters of the city, dating from the 17th century: 'I just knew that those documents were valuable and had to be saved.' 'I was met by a ball of fire and a cloud of smoke,' said Stapleton. Of course it helped that he was actually a member of the fire service. Reeling from the searing heat and gasping through the thick smoke, Stapleton crawled down the stairs of the city hall. From a first floor office he called Kilkenny station and then made his way down to the main gates of the building so the fire crews would have easy access. 'I did everything on reflex action, everything happened so quickly I didn't have time to think,' he said afterwards.

After the fire brigade had been alerted, Pat Slattery, in the city fire station then went about alerting the crews to muster at the station, as Kilkenny is a retained service. Within two minutes of the call being received, just after 6.25 pm, the first appliance was on its way to the blaze with five crew on board. Ultimately four units of Kilkenny Fire Service responded to the fire, and were able to do so almost immediately due to the close proximity of the station to the main street. Shortly before 6.30, the snorkel, which was to be so instrumental in bringing the fire under control, was on its way to the blaze. On the fireground, Kilkenny s/o Kennedy realised that they were dealing with something major, a fire in an elevated building, and it soon became apparent that the water supply available from the hydrants and the fire appliances could not supply enough to extinguish the blaze. When the fire brigade unit arrived from Freshford, the firefighters were told to pump water to the scene from the Nore river which flows through the city. As for the fire itself, according to local firefighters, until the later stages there was a lot of smoke visible but not much flame. However, when the fire reached full development it began to spurt through the tarred roof which had replaced the original tiles and flames also spat from the windows surrounding the clock. Firefighters Cleere and Lacey were on the snorkel, pouring water upon the blaze but, like the Officers and firefighters on the ground, they were also aware that the fire was in danger of spreading. 'That was one of the major challenges: Kilkenny is quite a compact city and the fire was, at one stage, threatening to spread to neighbouring buildings and, if that happened, there would obviously have been a far more serious incident facing the fire service,' says current ACFO John Collins.

Nevertheless, it was serious enough if you were beneath the clock tower which was ablaze. Four firefighters had entered the building in breathing apparatus. 'The biggest hazard was falling debris from above as the flames were raging over our heads,' remembers Joe Traynor. It was still less then an hour since the initial callout but already the battle was being won by the firecrews. Above on the snorkel, the firefighters had cleared most of the flame on the roof which had been threatening to spread to neighbouring buildings. In the clock tower, the fire was burning itself out and the four firefighters battling it from beneath had succeeded in clearing away material that might provide more fuel. It was approximately 19.15 when the fire was declared under control. Inside the hall, firefighters then began the process of extinguishing small spurts of flame and minor fires, all the while vigilant of falling debris from overhead. The fire was intense, but short lived and well within the capabilities of the six fire appliances and thirty-three firefighters who attended it. Today, the city hall has been beautifully restored, albeit with modifications after such a destructive fire. The lower part of the building was relatively undamaged although there was some water damage as a result of firefighting activities.

In the aftermath of this fire, which traumatised the city, the actions of firefighter Joe Stapleton were roundly applauded. Having served with the brigade in Kilkenny City for twenty years he was actually due to retire the Thursday following the blaze. 'His actions averted a possible multiple fire,' said Captain Hugh Corrigan of Kilkenny Fire Service. 'There was a danger the fire might spread to adjoining buildings but when we arrived Joe had everything at the ready and we were ready to get to work without delay. He deserves the highest level of praise.' Stapleton was not satisfied in raising the alarm either, he was determined to fight the fire alongside his colleagues, finally forced to retire due to sheer exhaustion. He was admitted to St Luke's hospital in the city to be treated for shock and minor injuries.

While this fire was a very visible — although not difficult — conflagration to fight, the fire which Kilkenny Fire Service was called to a little less then a year later was a truly bizarre one, which kept firefighters busy for over a week. And they never even saw a flame.

When the safety officer from Avonmore Creamery called the city fire brigade control room in Kilkenny to say there might be a fire in one of their milk silos, it would be the start of an eight-day firefighting riddle. Freshford were the first brigade to attend the incident, with crews from Kilkenny on the way. Essentially, what occurred was a 'hot-spot' — an intense smouldering — inside one of the dairy's milk powder silos. These are quite immense containers, this one towering seventy feet high. By 10.30 that morning, officers from Kilkenny had arrived on scene and the firefighters from Freshford had fire hoses trained upon the silo, which by now had become too hot to touch. Kilkenny also sent a fire appliance and along with the Freshford appliance, the firefighters sprayed the structure to try and keep it cool. The mystery was how to actually extinguish whatever was going on inside the silo.

What the firefighters needed, and didn't have, was a thermal imaging camera, a device which allows firefighters to see through solid materials, such as steel, and identify heat sources. In 1986, the technology was at a fairly rudimentary level but Dublin Fire Brigade was known to have one of the first of these devices in the country. ACFO Ryan called John L'Estrange of Dublin Fire Brigade and explained the situation. L'Estrange arrived with the thermal imaging camera that evening and they were able to identify the hot spots within the silo. The officers on scene hoped that by pouring water on the silo they would extinguish these hot spots, but the tactic hadn't paid any dividends so far. The same evening, Officer Corrigan, one of the county's senior firefighting personnel had arrived and instructed firefighters to vent the silo as there was a possibility of an explosion if the heat continued to rise at the level it was.

Firefighters then resumed pouring water on top of the steaming silo. So little progress had been made so far that the officers decided upon a rotating shift system to save the stamina of their men. Every brigade in the county would attend eight-hour shifts with ten firefighters per shift, overall 75 firefighters would be on scene. After three days of this policy, seven firefighters were instructed to go to the top of the silo and open it to check the temperature with a gauge. When the gauge was placed inside, the temperature was a workable 47 degrees, but six minutes later when it was checked again it had shot up to over 100 degrees. The silo was closed again and the fire hoses turned back on. In all, over the eight days of this operation, over 350 million gallons of water were used. To satisfy such an immense demand the fire service once again had to resort to nature's water supply at a nearby river. After eight consistent, tedious days of hosing the silo the 'hot spots' inside the container were seen to have finally dissipated. When firefighters opened the silo to check inside, the vast quantities of milk powder inside had been solidified by the intense heat.

Going Underground

On 6 February 1973, fire brigade units from Tipperary and Kilkenny responded to a fire at the Ballingarry coal mine in Co. Tipperary.

Earlier that morning, Dick Ivers, who was working at the pit had smelt smoke and had alerted the ground staff who discovered fire in the mine quite close to where some maintenance men were working below. Five men, Paddy Tobin, George Ivers, Joe Tobin, Ned Grant and Martin Devitt, made their way to the surface via a two-mile-long airshaft before travelling back down the main mine shaft again to rescue their colleagues. Shortly afterwards, eight other men made their way to the surface. 'We knew there was an air shaft and we made for it (when we realised there was a fire) and got out. We got a bit of a shock.'

There were, however, four men still trapped in the increasingly dangerous environment, including Dick Ivers, the man who had first noticed the fire. When firefighters arrived, the flames were threatening to surround the trapped men. Conditions in the mine were difficult and very basic since it had been closed for business — the only reason the men had been working there was to keep the machinery in working order so the mine could be sold on. It was also completely unknown what had started the fire, when Ivers alerted the pit, he had seen smoke but the exact source of the blaze was unknown. Subsequently, it was blamed on an electrical fault.

Eight Brigades from throughout Tipperary and Kilkenny, in addition to ambulances, were dispatched to the scene. It must be remembered that this was a very different time for the brigades, almost a decade before legislation that would modernise the service. Nevertheless, the firefighters were there and preparing to go underground in heavy breathing apparatus to stop the fire and rescue the four men. An added danger was the structural integrity of the mine itself. 'There had been a (partial) structural collapse and for some time it was feared that the force of the water used to quell the flames could cause a serious cave-in,' said Clonmel fire officer, Brian McMahon.

One of the firefighters who was there that day and has served over thirty years with the Brigade in Kilkenny was Joe Traynor, today an s/o in that city. 'What I remember from the day was dirt and fatigue.' For any firefighter, full-time or retained, this was a strange and alien fireground. 'I remember it was very tiring going down and coming back up in the breathing apparatus, it really took it out of us.'

The policy was to go down and fight the fire, come back up and rest while the relief crew took over; then go down again. 'Obviously, firefighting in confined spaces wasn't part of our training back then, but we got on with it. Sheer willpower really.' It needs to be remembered that only a short time before this incident, these firefighters were either engaged in their other, full-time, jobs or were looking forward to another normal day, now they were suited up and ready to go down a mine shaft. That's the nature of the retained service.

Below in the shaft, an unpredictable fire was burning. 'It wasn't one fire really, but a series of fires that were burning behind the coal seams themselves, very hard to get at.' In addition to the dark, the fatigue, the smoke, the flames and the complication of four men sitting in an air pocket waiting to be rescued, the firefighters also came across another problem. 'Rats, big ones,' says Joe Traynor. 'We knew there were rats down there: some people have a problem with them, others don't. But it's different when they start to bite. When I was going back down into the mine on another firefighting sortie, I saw a firefighter being helped up by another colleague. He was clutching his hand and I could see it was bleeding quite badly. He had just got quite a nasty bite from a rat.'

In full heavy-breathing apparatus, the firefighters fought the blaze for well over four hours, finally clearing enough of the smoke to effect a safe rescue of the men and have them brought to the surface where relieved family and colleagues were waiting. Dick Ivers, Tony Butler, Paddy Moriarty and Willie Cleare were remarkably calm when they emerged from the mine according to newspaper reports. 'We were not unduly worried because we were in contact with the men above all the time and we knew what was going on, we just sat tight in the air pocket until the fire was out and they came to get us,' said Ivers, who had a quarter of a century of mining experience behind him. The men had been trapped underground for six hours.

While the situation had not reached a perilous condition for the men below ground, this was likely only due to the swift actions of their colleagues and subsequently, the firefighters. 'It's one of the fires that's still remembered here because we don't have many mines in this country, a lot less now and when something like a fire happens in a mine it can be very serious,' says Joe Traynor. For the retained brigades, the operation was a great success. 'To be honest, nobody had any prior experience of that sort of thing. It did have unusual and uncomfortable elements in it, dark, lots of smoke and a fire that was very difficult to find, but we did get in there and do it. It was definitely an experience and when you remember the fires and incidents in your career, that's definitely one of them.'

Stretched to the Limit

It was lunchtime on 7 January 1992 when the fire began to smoulder amongst the piles of cardboard and packaging deep within the dry goods store. Within hours, the paint on the exterior of the Halal meat processing plant in Ballaghadereen, Co. Roscommon, began to blister and peel, the intense heat consuming vast quantities of dry toxic fuels. For the retained fire service in Roscommon, this would present a huge challenge. To successfully contain, suppress and eventually extinguish a fire on this scale would require logistics and on-scene presence that the retained fire service is simply not designed to cope with. Firefighters would eventually be on scene for ten straight days, extinguishing a pernicious and persistent blaze.

The fire at the Halal Meats plant has been debated at both local and national level. Was it an 'insurance job'? Was it an electrical fault? The causes are still debated by the firefighting fraternity in Roscommon, but they are secondary to the fact that the job was done successfully.

The alarm for the fire went off just after lunchtime on a bright, cold winter's day. 'It was just after half-one when I got the call,' remembers Ballaghadereen s/o Fergus Frain. 'I could see the smoke and I called Castlerea fire station, the nearest, for backup, before we left the station.' Like the fire at Cow Comforts in Belmullet, the Halal Meats plant was a large employer in a town with a small retained brigade, consisting of eight men including the s/o and SUB/OFF. Local firefighters arrived on scene at 13.50, with Castlerea arriving closely after at 14.15.

The scene which Fergus describes that day, while maybe not as apocalyptic in initial appearance as the one in Belmullet just over seven years later, was fairly chaotic. 'We had a rough idea of the layout of the structure and when we arrived we knew that it was the dry goods store which was up in flames. It was the worst possible place for a fire to have started as that was where all the cardboard and plastics needed for the packaging phase of the production process was stored. The area itself was about forty metres by ten metres and like the rest of the building was enclosed by cladded walls. The problem was, there was only one door in.' Thick black smoke was beginning to billow out of the building at this stage; the flames were low-level but burning strongly. 'That part of the building only had one door, so with an environment like that it was completely out of the question to allow any firefighters to enter.' The flames were moving quickly, devouring the interior of the dry goods area with frightening speed.

Fergus Frain remembers the smell of burning plastic and a lot of . thick black smoke. 'The air was thick with dank smoke,' remembers ACFO Frank Cafferty. 'I remember running my hand across the surface of one of the vehicles, and the visor of my BA set. Both were thick with grease.' As the fire began to spread, a series of decisions needed to be made. The biggest problem on the fireground was access. 'What we had here,' says Frank Cafferty, 'was a relatively simple building. However even very simple structures can become hazardous places when fire is present.'

Within the space of forty minutes, the area was completely gutted. Roscommon CFO Cathal McConn was fairly sure that there was no-one inside, which meant that the scene was one of calm urgency rather then desperation. The dry goods store was a holding area where there generally wouldn't have been anyone working, except for a forklift operator on occasion. 'I did know that there was plenty of combustible material there though, particularly in the shape of flat-packed boxes used for packaging the meat.' The environment had been judged as too hostile for firefighters to enter, but all the men remember that the biggest problem in this case was access. 'It was a simple structure, we knew there was a large production area with a cold and dry goods store off it,' remembers Fergus Frain. 'However, we knew that parts of the cold goods store (measuring ninety metres by forty metres) were automated and we weren't familiar with the layout. To get into that cold store was an option but one that was too dangerous. With the doorway to the dry goods store swamped with thick smoke and flames, the only way into the dry goods store was through the main building itself.'

When the factory was operational, the goods from the dry goods store would be removed by forklift and transported to the first floor of the production area, the largest part of the building. The boxes were then assembled and dropped down through chutes onto the main production area. The meat was vacuum packed into the boxes here, blast frozen and then brought into the cold store. The cold store itself was automated with one aisle for access where pallets could be stacked on a rack: when one rack was full it would automatically move forward and make space for the next one. 'That was one of the major challenges for us; there was only one door to access the cold store so there were very limited points of entry. It was hard to ascertain what exactly was going on. We were to discover that this fire was low intensity, in that the flames weren't going through the roof, but that there was a lot of smoke,' says Cathal McConn.

'The cladding on the exterior of the cold store was melting and peeling, you could feel the blistering heat,' adds Fergus Frain. 'So we knew that while the flames might have been low, thick black smoke and plenty of fuel made it clear that it wouldn't be an easy fire to tackle, exacerbated by the fact that we couldn't get men inside — it was just too dangerous.'

The fire had also penetrated the void (insulation panels) between the cold store and the dry goods store. These panels were held together by nylon bolts. There were no metal bolts used because they would have acted as a heat conductor into the cold store, creating hot spots which could have spoiled meat. Of course nylon bolts couldn't withstand the heat that was being generated, and when the temperature rose they began to melt and break. The insulation panels collapsed exposing polyurethane foam, more toxic fuel for the flames. Nothing in that area should burn easily, but with temperatures as high as they had become, anything would burn. Thriving in the vacuum between the cold goods store and the dry goods store, the flames fanned upwards into the roofspace above the cold store and then gradually spread. As the flames spread slowly the paint on the exterior of the cold store roof could be seen to blister and peel, another indication of the intense heat generated by the slow-burning flames. 'We knew that the flames had got in between the inner and outer walls, so it was going to be hard to tackle because we didn't know exactly where the flames were and more importantly where they could get to,' says Frank Cafferty.

As the fire took hold, the scene surrounding it became increasingly chaotic. In addition to the gathering crowds, conflicting advice from people who weren't trained in firefighting added to the tension.

The benefits of using fire-retardant foam are well detailed, but for it to be effective, you need to have enough of it to smother the fireground completely.

'The main problem was that we were being told by people who weren't firefighters that we should hit the fire with foam.' 'We worked out how much foam we would actually need and we would have needed somewhere in the region of five thousand gallons. We couldn't have got that much from anywhere.' Fergus Frain also remembers the personnel from the insurance company in London. They were conducting their own investigation into the cause of the fire and they had wrongly come to the conclusion that the fire had started in a pallet of flat-packed cardboard within the dry goods store. 'I remember on the third day of the fire they asked to speak with me as the s/o of Ballaghadereen Fire Brigade. They decided to do a field test to illustrate their theory of how the blaze started, that it had been sparked by a cigarette, a theory we didn't agree with. Of course they were wrong.'

In addition to unsolicited bad advice and a volatile fire, the crews, and particularly the CFO, were also aware of the strain that the Halal blaze was putting on Roscommon's firefighting resources. If another major fire was to take place elsewhere in the county, they would need resources ready. 'On any big fire, which is going to put such a massive burden on the firefighting resources of the county, you have to have a contingency plan,' says Cathal McConn. 'There needs to be backup in case there are other callouts in the county that have to be attended. What we had to do was to rotate the crews from throughout the county. Once any unit had clocked a certain amount of on-scene time they were sent home for rest. You then call in another unit. That is your basic plan. You have to keep the men fresh and fit for work. But even when you're at home, you can't really relax when you're dealing with something like that. You might get two hours in bed and a shower but the phone would be ringing constantly, often with media calls. Then a couple of hours later you'd be back up and out on scene again.' As the blaze entered subsequent days, it put an inordinate amount of strain on the lives of the firefighters involved.

On the fireground, the huge logistical requirements of the firefight were being hampered by public interest. 'We had a big problem trying to keep members of the public out of the way to enable us to do our job effectively. This was a fireground after all,' says Cathal McConn. 'We were presented with quite sizeable logistical problems, one of which was the vast amounts of water arriving on scene in milk tankers provided by locals. Difficulties arose in getting this water to our appliances. Essentially, we needed to lift pumps by forklift up on top of these trucks and pump the water out of them or to construct temporary lagoons.

'Availability of water was never really a major problem as we had the supplies from the tankers and local sources such as rivers and streams. Our major challenge was getting water onto the seat of the fire itself. When the cladding walls of the dry store collapsed, they collapsed inwards over the seat of the fire. We were hitting the fire with thousands of gallons of water but a lot of it was actually deflected by the fallen cladding, which was effectively shielding the blaze.'

A fire of this scale obviously generated media interest, and while the firefighters have generally positive memories of the media's behaviour during the course of the fire, taking care of media enquiries meant yet more work for a brigade that was not used to this level of attention.

'You have to remember that it wasn't an easy story for the press either. Particularly for the television crews because the fire was all happening on the interior of the building. There were no flames shooting through the roof or anything. It was a persistent, low-intensity fire. Therefore, the only way we could really see it was from the snorkels or platforms which we positioned around the fire. The snorkels meant that we could hit the fire from above but it also meant that we could show the media what exactly was going on. We weren't trying to create a great impression of the work we were doing, but we did want to give them the real story of the fire and that was only apparent when a camera could get up in the air over the building and get a look at it. That is how you should handle the press, I think: if you can give them a good story without disturbing your own work on scene then you should do it. If you handle the media right, they will respect you for it,' says Frank Cafferty.

The big problem for firefighters was finding suitable access points to the building. While the dry goods store was deemed too dangerous to enter, there was a huge problem in getting fire hoses to hit the seat of the fire itself. 'There was a distance of approximately seventy metres from the only entry point that was open to us — at the far end of the building — to the seat of the fire itself. We couldn't get men in close enough to hit the fire head on,' remembers Cathal McConn. 'One of the options open to us was to create our own access point. That meant ripping open a section of the building. The owner of the building was particularly keen on this, recommending that we enter a section of the cold store after breaching the wall. It was something that we thought about, but we didn't know exactly what was going on inside the building, It was too dangerous.'

Inside the factory, an increasingly hostile environment was being created by the fire. Intense heat, thick smoke from dry goods and frozen meat meant that visibility was virtually zero or a few metres at best. There were other dangers too. 'Throughout the course of the fire we could hear very unusual high pitched noises. A bang followed by a crash.' Firefighters heard a large number of them. For a considerable time the cause was unknown. 'We learned that the cold store used freon gas, and that it had a number of evaporators suspended from the ceiling,' explains Fergus Frain. 'Each of them was about the size of a large washing machine. As the structure began to buckle from the intense heat, the nylon bolts holding these evaporators in place melted and broke, with the evaporators crashing onto the floor below.' If firefighters did enter the building, they would be stepping into an atmosphere choked with smoke, intense heat, limited visibility and at each step they were in peril of being crushed by these collapsing evaporators. The fireground itself was no environment for a firefighter to be entering into but fire officers did deploy firecrews into the main production area, where the fire hadn't taken hold. This was purely as a defensive line should the flames break through the wall. 'It was unlikely that the fire would have broken through that wall, as it was the only solid wall in the building,' says Cathal McConn. 'However, we needed firefighters there because the wall itself didn't go up to the roof. There was a chance that the fire could have got through near the first floor. It was the only place within the overall development that we could have deployed firefighters in safety. It was simply too dangerous in either the cold store or the dry goods store.'

While a mammoth task in terms of sheer firefighting resources, one of the positives about the Halal fire was that there was virtually no risk of collateral damage. The fire had nowhere to spread to. The fire crews were able to go about their work knowing that while the cold store part of the premises itself was doomed, the rest of the factory was relatively safe. Fergus Frain remembers that once this became apparent, there was absolutely no point in risking people when the fire was under control. 'We couldn't touch the parts of the building that were burning, but we knew it wasn't going to go anywhere else. We deployed the defensive line of firefighters just in case but the fire was contained. The worst-case scenario had become apparent quite early on in this case. There was no panic. We knew that we would be on scene for as long as it took us to totally extinguish the fire. So we just got on with it.'

Inside Halal Meats, vast quantities of beef had been destroyed by the fire which at this stage had totally consumed the cold store. The smell of burning meat hung thick in the air. Once the building had risen in temperature by five degrees, the Department of Agriculture effectively wrote off all the produce inside. There were nearly 100 articulated lorries parked from the factory into the town of Ballaghadereen itself, all waiting to take out the scrapped meat. The Department was well prepared and had the trucks there waiting. It was a situation where everyone was well prepared to respond, even taking into account the difficulties of dealing with a fire on this scale.

For firefighters, their lives became severely disjointed over the days of the blaze. 'I remember my wife sending my daughter down to the factory to plead with me to give a statement to the papers,' says Fergus Frain. 'Since I'm a station officer, my home phone is essentially a 999 line, so it was ringing constantly at all hours of the day and night. That was hard on the family because I wasn't there, they didn't know what was going on — all they knew was that there was a huge fire and I was there fighting it. It wasn't fair on her or the children.'

A fire of this scale has a cultural impact on a fire service like Roscommon's. Everyone in the service at the time will either have been on scene or covering for someone who was on scene. 'While it was quite harsh being there for such a period of time, I do remember a terrific camaraderie amongst all those involved. I remember an officer from Sligo, of senior rank to me, asking was there anything I needed help with. When I told him we were having problems with a pump at the far end of the building, away he went to get it fixed, no questions asked — that sort of teamwork was great,' adds Frank Cafferty.

The last firefighter walked away from the gutted building almost a week after it initially caught fire. Once the blaze had been brought under control, it remained difficult to extinguish entirely, with days of dampening down flare-ups and hot material remaining. This was a hugely complex fire to deal with, both in terms of firefighting and also managing available resources and deploying logistics properly. 'Before I retired I was a firefighter for almost forty years, between here and the UK and I've come to know that many people think that firefighting is simple,' says Frank Cafferty. 'Well, it isn't and this fire was yet another case of that. I may be retired but I retired without knowing everything there is to know about firefighting, and I don't think anyone ever will.'

Part Three

Famous Fires

Trinity College

One of the most iconic settings in Dublin city, Trinity College is a historical, architectural and cultural jewel. Its collection of magnificent buildings contains treasured artefacts which chronicle the history of a nation. However, in July 1984, when a serious fire took hold in the Dining Hall of the college, Dublin firefighters and college staff found themselves in a battle to save the building and its famous contents.

The night of Friday the 13th, bad fortune's iconic date, was when the blaze was noticed first by Professor George Dawson, who was in the nearby Senior Common Room and then noticed flames and smoke arcing from the bell-tower above the dining room. 'I frantically smashed the glass of the fire alarm and contacted the college porters to tell them to call the brigade.' At the same time nearby on the campus, Mick McCaughan, the college's entertainment officer was making dinner for friends when they noticed the smell of burning. Along with his friend, Barry Cooke, both men ran to the front gate to raise the alarm. Meeting Professor Dawson, they then made their way to the building to salvage what they could from its hallowed halls.

The walls of the building's interior hosted huge portraits of former provosts and benefactors of the college. On arrival, the students, their colleagues and the professor began to organise a salvage operation to rescue the Dining Hall's treasures. They emptied three fire extinguishers, to no great effect, on the flames before forming a human chain, numbers boosted by other students and college staff, to evacuate what they could from the fire's path. The rescue effort was largely a successful one, retrieving most of the building's treasures. Mick McCaughan remembers the scene: 'We scrambled to gather what we could. Paintings that you were once frightened to touch were being pulled frantically off the walls. The hierarchy was forgotten and whether you were a student, a passer-by or a professor, it didn't matter. Everyone was reduced to a pair of hands. It was very strange. It was a moment of chaos in a place which had always been so dignified.' Lost to the flames were several old portraits of various Princes of Wales and Church of Ireland bishops which they were unable to take from the walls. In addition to paintings, chairs, silver, lamps and tables were also rescued. Firefighters eventually had to stop any more forays into the building when the blaze took hold of the supporting beams in the roof.

Nearby, Dublin Fire Brigade dispatched four firefighting units to battle the blaze. The tower, with its famous clock, was swathed in flame when they arrived. The flames had taken hold of the loft and smoke was flowing freely from the roof slates. When one of the firefighters was raised up on the extended ladder over the flaming roof, he pulled the lever on his hose, expecting to hit the flames with a torrent of water. Instead, a trickle of water dribbled sporadically from the nozzle. The ancient pipe system in the buildings was too small for the hoses. Without large enough water mains to which to attach the fire hoses, and hence sufficient water, the fire would burn out of control very quickly.

By 7.30 pm the fire had taken hold of the left-hand side of the roof, and slates, which were splitting from the heat, began to slide to the courtyard below. In their place, flames leapt through the holes. Fanned by a strong south-westerly breeze, the fire then proceeded towards the centre of the roof. The nearest working hydrants were located on Pearse Street and were put to use: the street itself was sealed off and water began pouring onto the flames.

In the interim period, during the frantic search for water, firefighters had donned BA sets and used fire axes to smash through a laundry in the basement of the Dining Hall while their colleagues above made a frontal assault on the fireground through the main door. The wooden interior of the building was filling with smoke from the flames which were ravaging the roof above. Once water had been sourced, it began to flow freely onto the flames from the latter above the roof, but considerable damage had already been done. By 8.30 pm, flames were spurting several metres from the roof of the building, which had been largely destroyed. At this stage, firefighters were also concerned about a neighbouring building catching fire and a third fire appliance arrived to protect it.

Then, a crashing noise was heard from above as the building's famous clock tower crashed violently into the east gable. Melting the lead in the roof, splintering slates and smashing beams, the fire spread backwards, destroying the entire roof in the process. One of the wooden beams above, devoured by flame, crashed to the floor below, striking one of the firefighters on the back of the neck. He was taken to hospital although his injuries were not too serious.

Outside, at College Green, a crowd of thousands had gathered to watch the drama unfold in a mixture of horror and fascination. An *Irish Times* reporter said, 'It was like half going to the movies, half going to a wake.' After hours of fighting the flames from outside, inside and above the hall, it was finally brought under control. The Dining Hall was structurally intact but the damage to it was severe. The panelled walls, upon which a few surviving paintings miraculously still hung, now looked down upon a chaotic mix of burning ember, masonry, wood and water.

Damaged as it was, the Dining Hall was totally rebuilt at a cost of £2 million (old Irish) pounds. Some years prior to the blaze, a detailed survey of Trinity's buildings had been carried out and its detailed drawings and accurate measurements of every aspect of the hall proved their worth when it came to restoring it to its former glory. As for its furnishings and fittings? The vast majority of the paintings had been rescued, save for a few casualties when the roof collapsed. Smoke-damaged and shrunken canvases were carefully repaired by Welsh craftsmen while ornate frames were reguilded in Dublin and the grand chandelier was also restored to full health. Even the clock, notoriously stuck at a few minutes to 12 before the fire, now told the right time. Soon, Dublin's great hub of learning returned once more to its normal rhythms. For the city, it was one more night now only remembered in newsprint unless you were there. For men like Professor George Dawson, pictured clutching a blood-stained napkin to a hand wound sustained while smashing the glass on the fire alarm, the date of Friday the 13th would remain ominous ever more.

Powerscourt Fire

A decade before the Richard-Cassells-designed Dining Hall in Trinity was gutted by fire, another iconic building designed by the same man was destroyed by a devastating blaze. Powerscourt House in Wicklow erupted in flames in the early hours of 4 November 1974, just days before the house, a magnificent 18th-century mansion, was due to be opened to the public for the first time. Indeed there are almost supernatural overtones to the fire in this glorious house: before the area where the house was built became known as Powerscourt, after its builder Eustace la Poer, it had been known as 'Baile na Tinne', which translates directly as the Town of Fires.

Constructed in the 1730s, the house was a thing of fancy for the viscounts of Powerscourt, particularly Mervin, the 7th Viscount Powerscourt who inherited the house in 1844 at the age of thirteen. Eight years later, when he became twenty-one, he set about flamboyantly spending his inheritance. He made the house bigger and constructed a huge ornate Italian garden towards the front. From the finest craftsmen throughout Europe he scavenged medieval gates, monumental statues, ornate fittings and baroque furniture. He developed the house into a truly spectacular, if somewhat haphazard, collection of styles.

The house passed from the Powerscourt family in 1961 to the Slazenger family, of tennis equipment fame, for the sum of 300,000 Irish pounds. The house had remained enigmatically reclusive from the local townspeople since it had been built over 240 years previously but now, Ralph and Gwendoline Slazenger began plans to have it opened to the public.

The night of the fire was non-descript until the lights suddenly darkened as Gwendoline Slazenger was reading in her bedroom. 'I didn't take much notice of this as and thought it was just the electricity acting up again,' she told the press the next day. 'Then I heard a tiny tinkle. We had installed a bell in my mother's room but since my mother wasn't there, I decided to investigate and woke my husband.' When Ralph Slazenger went upstairs and noticed electrical sparks in one of the bedrooms. Gwendoline ran and called the fire brigade.

Forty-five firefighters were despatched from four stations as the fire began to take hold at the old mansion. Within twenty minutes of sparks first being noticed, Mr Slazenger said that the top of the house was ablaze. The family believed that the fire originated during the previous afternoon when a chimney fire had not been properly extinguished. When the firefighters arrived, they were faced with a challenging situation. Because the blaze was situated high in the building it would obviously be more difficult to access and there were also the added dangers of the floor and roof collapsing in on top of the fire. If the fire hadn't engulfed the top floor there may have been a slight chance of stopping it, as the house's firefighting equipment had been stored there, however, like much else, it was completely destroyed in the blaze.

Under Chief Fire Officer Reginald Sharpe, firefighters conducted a quick salvage mission to save as many of the house's precious items as possible before they were destroyed by flame. However, the smoke and flame quickly became too thick to penetrate and the flames, aided by strong winds, began to move quickly through the old house. It quickly became evident that the centre of the house was doomed so an alternative firefighting and rescue plan was quickly devised. The fire appliances' water was concentrated on the two wing sections of the house. Pumping at a rate of 2,500 gallons per minute from the nearby ornamental pond, the hoses soon had an impact on the flames which were licking the walls on either wing. However, by this stage, the centre of the house was a raging inferno. The efforts of firefighters did pay dividends as the wings of the house, particularly the east, survived the blaze without very serious burning, albeit with heavy water damage.

The current editor of the *Irish Times*, Geraldine Kennedy, was on scene as a journalist when she spoke to the shell-shocked owners of the house. Both Ralph and Gwendoline Slazenger had spent the months and years leading up to the tragedy engrossed in their task of turning the house into a major tourist attraction. Gwendoline said, 'The house was looking at its most gorgeous only last weekend. We spent £900 on roof repairs last week in final preparation for making the house public and the bill isn't even paid yet.' Nearby, their precious roof had collapsed inwards upon the blaze.

The pride of the house, the eighty-by-forty-foot Great Hall, was built to entertain King George iv of England during his 1821 visit to Ireland. After the fire it was reduced to a charred, blackened shell. Its priceless contents included numerous oil paintings thought irreplaceable, Waterford chandeliers, hand-woven carpets and rare mirrors. The fire also claimed the Ballroom and Long Gallery with many priceless works of art, The Tudor Hall, The Leisure Room, The Dining Room, The Cedar Room, The Fan Room, the Boudoir and the five bedrooms and bathrooms on the top floor. When the smoke settled on the morning of 4 November, the vast majority of Powerscourt house was effectively destroyed. Chief Fire Officer Sharpe said that the fire had spread with such devastating rapidity that it became apparent that the only course of action which was open to his firefighters was to try and salvage as much of the house and its possessions as possible. The pictures published in the press during the following days conveyed the extent of the devastation. The Slazengers at the time had no plans to rebuild as they walked through the ashes of their former home, still populated by firefighters dampening down the embers. 'The burning of its contents is not only a personal loss but an immeasurable loss to the nation,' they said. Gardaí were satisfied that the fire was not the result of anything except a terrible accident.

In the aftermath, the Slazenger's stoicism gave way to some dry wit.

Taking issue with some inaccuracies in *The Irish Times* report of the fire, Mrs Slazenger had a brief letter published in the paper on 8 November:

> To introduce a note of humour into this disastrous situation, I would like, being a woman, to correct two small items in Geraldine Kennedy's report (5 November). I am in my early fifties, not sixties as she reported, though no doubt after being up all night running back and forth through smoke I may have looked 60. Also, I did not go to wake my husband in his room. He was sleeping by my side as he always does. In these tragic circumstances, these little things are important to a woman.
>
> Yours etc.
>
> Mrs G.M. Slazenger

Slane Castle

Lord Henry Mountcharles stood ashen-faced outside the smouldering ruins of Slane Castle early on the morning of 19 November 1991. 'It is not what I have lost, it is what the country has lost,' he said as behind him the once magnificent fortress lay scorched and ruined under a pall of smoke. Nurtured by strong winds, the fire which had started in the basement of the castle raced upwards, destroying much of the building in a furious blaze.

For Mountcharles, two centuries of continued ownership of the castle by his family had been blighted in the most terrible fashion. The castle, as famous for the annual massive rock events held there as it ever was for its splendid architecture, had been devastated. His raw emotion was, however, tinged with a hint of strong determination: 'It's terrible to see it just like a shell,' he said, 'I'll find some way to rebuild it.'

For firefighters, the night of 19 November would see them participate in one of the most dreadful battles which the castle had ever known. Built as a fortress, it was designed to keep people out, that night firefighters would have to storm the castle itself to fight the flames.

It was around 4 am that the fire in the basement kitchen of the castle began to take hold. Andrew Cullen, an employee at the castle, who was sleeping on the premises, noticed the smoke and immediately called the emergency services. The east wing of the castle was ablaze by the time firefighters arrived. Lord Henry Mountcharles lived in his other home nearby in Beaupark. One his way to the fire he remembers seeing the plume of smoke and flames in the distance. Speaking to the *Irish Independent* the following day: 'As I drove over the Boyne bridge I saw flames shooting from the castle. When I arrived the east wing was already ablaze. It had just swept through the upper part of the house.' Inside the castle, the precious treasures of the east wing were being consumed.

Michael Stack, CFO with Meath Fire Services, remembers the scene when the brigade arrived. Throughout the course of the fire, firefighters from stations in Navan, Drogheda, Dunshaughlin, Kells, Trim and Nobber would battle the blaze. 'It is a castle, designed to be impregnable, to keep people out, so access was a major problem. We couldn't get our fire engines around the back.' The steep slope to the rear of the east wing meant that the only passage open to firefighters was through the front of the building where, along with Henry Mountcharles, several fire officers struggled to salvage what they could as the flames advanced rapidly through the building. Assistant Chief Fire Officer John Moore remembers seeing Mountcharles running from the castle holding armfuls of its irreplaceable treasures. 'He was well organised throughout the fire. He was taking his paintings and family heirlooms to safety. I remember him carrying swords, fighting swords, which I imagine may have been used in battles long ago.' The castle's owner pinpointed items to the firefighters which he wanted saved, his coolness in the intensity of the situation was not lost on one of the firefighters, who said in the aftermath of the fire, 'He's a very cool man, he has a lot of class.'

It was a race against time which could never be won, while some of the castle's most treasured artefacts were indeed saved, many others were lost and the men had to pull out as the heat intensified and smoke billowed through the halls. 'We managed to save some (items) but smoke and flames became too much and we feared the roof would fall in.' Mountcharles was later taken to hospital suffering from the effects of smoke inhalation but was released soon afterwards. One of the castle's most precious items, a portrait of George IV by Thomas Lawrence, was rescued.

The battle which raged around Slane Castle was now a siege of surreal qualities. Ladders braced the walls and hoses snaked from the nearby Boyne River as firefighters directed gallons upon gallons of water on the flames. 'The fire was spreading very rapidly,' remembers

Michael Stack. 'The flames spread underneath the roof spaces and between the floors. There were a lot of latticed walls with strips of wood behind the plaster and, when the timber started to burn, we couldn't get at it to extinguish the flames.' However, another firefighter remarked that the castle's ancient construction methods actually contributed to its structural survival. Commenting on the oak rafters, he said that if the rafters had been iron the entire structure would have collapsed with the heat.'

Some of the most striking images of the devastated castle are from above, showing the terribly scarred roof of the fortress. It was simply too high for firefighters to gain access to, and the flames destroyed the second floor, which now looked upwards through a ruined roof.

Residents of Slane and the surrounding area gathered around as the flames reached their denouement. The Castle was, and remains, one of the area's major attractions and employers. Some stood silently and watched as flaming embers flickered into the night sky and wooden beams came crashing down noisily through the ruined interior. Farm manager Gus Doggett and his wife Marian broke down in tears as they witnessed the inferno. 'It would break your heart,' said Marian. 'For hundreds of years the castle has been the focal point for the whole community. Many people have made their living from it, it's a terrible tragedy.'

For firefighters, five long hours of fighting the flames were coming to an end. Around 9 am, the fire was thought finally to be under control. The amount of water which they had thrown at the blaze meant that what was not burned or still aflame was now soaked in water, which streamed down the walls and parapets of the castle. The fire had been successfully contained in as much as it could have been considering that flames had already taken considerable hold in the short time it took for firefighters to arrive.

The eastern portion of the castle had borne the brunt of the fire's worst ravages: Fanning upwards from the basement the flames had started a mushrooming effect which enveloped the east of the building, including the castle's ballroom and its ornate domed ceiling. The amount of water which had to be thrown upon the fire was now causing more problems. Areas of the castle had become swamped in water and there was danger of further structural collapse as it pooled in certain areas. Meath Fire Services came in for some criticism from some quarters for not doing enough to counteract this. This was particularly the case in the ruined ballroom toward the rear of the castle. However, Michael Stack said that water was pumped everywhere in an effort to stem the blaze and a committed effort was made by firefighters to do whatever could be done to stop flooding in certain parts of the castle.

In the aftermath of the inferno, which was still smouldering into the afternoon of 19 November, the grim task of assessing the scale of the devastation could begin. For Henry Mountcharles, it was the first day of the castle's rebirth and over the following years, through ingenuity, determination and skilled craftsmanship, much of the castle was restored. Today, it still stands resplendent on the banks of the Boyne.

The Irish Times

For news reporters, it's a strange day indeed when the news actually happens in your building. Such was the case on 17 September 1951, a relatively normal day in *The Irish Times* offices until staff reporter Noel Conway, sitting in front of his typewriter, heard the scream of 'fire' from outside the window. Looking out over Fleet Street he saw the glow of flames licking upwards: 'Almost immediately, this glow seemed to grow into a raging ball of fire.'

The terrible fire which gutted *The Irish Times* building that day, destroying the entire building from the ground up, was exceptional for two reasons: one, that despite huge sections of the building collapsing and the burning of vast quantities of highly flammable products, nobody was killed or seriously injured and two, that despite this devastation to its headquarters, the paper was still on sale in the shops the following day, the destruction of its own offices the main headline.

When the fire broke out at the landmark building, assistance from the fire brigade was literally around the corner. Five sections of the brigade were sent straight away, and once they arrived the scale of the fight on their hands became apparent. The blaze had started in the machinery and dispatch rooms, the result of an electrical fault. These rooms, the technical heart of the organisation, housed vast printing, lithographic, engraving and binding machines in addition to racks upon racks of paper, inks and oils. Once the flames began to feed hungrily on a fuel source as flammable as this, they spread relentlessly, sending massive sheets of flame into the sky.

Inside, speed was of the essence. Mr B. Rourke and Vincent McCormick were working in the case room, close to the seat of the fire. When they saw the rapid advance of the flames, they didn't even have time to grab their coats. Likewise for news reporters. Although they were housed in an adjoining building, the intensity of the blaze generated a searing level of heat which meant that the section had to be evacuated very early in the fire. At this stage, the battle was on to save what could be of the buildings. The threatened section of the Fleet Street building was thick with firefighters and fire hoses, pouring thousands of gallons of water onto the blaze, but they were unable to get near the seat of the fire, in the machinery and dispatch room. The roof of this section, buckling with the tremendous heat, collapsed soon afterwards, sending billows of thick toxic smoke out onto the nearby streets.

Outside, as was to be expected, massive crowds were gathering to survey the destruction and the firefighters in action. Gardaí were deployed in large numbers to keep the crowds back and it is worth noting that this fire was the first in which gardaí used hand-held radios to communicate. The sheer numbers which did gather to witness the inferno did little to help the firefighters and indeed they hindered, unwittingly, the deployment of ladders and firehoses to get at the fire.

Back on the fireground, the salvation of the building's contents was now paramount. What machinery could be saved had to be saved: it was incredibly expensive and very difficult to replace. Equipped with their traditional fire axes, the firefighters began to hack their way into any section of the burning building where they could gain access; dragging their heavy fire hoses behind them they got to work on saving what they could, most notably a recently acquired American rotary machine. In areas of the building as yet unaffected by the blaze, what remained of *The Irish Times* staff salvaged what they could in terms of files, folders, archival material and office equipment. While machines and equipment could be replaced, the most tragic loss to the city and the country that day was the destruction of back issues of the newspaper itself. They were irreplaceable, each one similar, in a historical sense, to losing a day in history.

By nightfall, Dublin Fire Brigade has succeeded in bringing the fire under control and salvaging what they could. Nevertheless, they remained on scene to control the fire all through the night and into the following day when, to their surprise as well as everyone else's, *The Irish Times* of 18 September 1951, albeit at a slim four pages, went on sale. The Trojan efforts of the firefighters that day were replicated by the paper's staff, who resolved that the show must go on and that *The Irish Times* must put out an edition. Despite the fact that one third of their offices were being consumed by flame, the commercial section of the paper continued to operate, booking advertising space for the next day. A huge debt of gratitude was owed to the now-defunct *Evening Mail* for helping their press colleagues put the paper together, though at four pages it was more a symbolic edition then a standard one. While there is no record of any serious injury or casualty during this epic blaze, *The Irish Times* set the story straight by revealing that in fact their were four lives lost. In a truly sad story, Matilde, the pet cat of the Circulation department had lost her entire family, four kittens, in the fire.

The following day, the afflicted areas of *The Irish Times* building consisted of sopping wet charred timbers and the detritus of a destroyed office. Here and there, blackened filing cabinets stood, their contents a mound of charred ash. One inspirational snippet from the destroyed archives which survived was from a speech delivered by a member of Dublin Corporation. It simply read, 'There's no panic.'

Another inspiring story to come from this day of terrible destruction was that of Kieran Fagan. Only six years old on the day of the fire, Kieran was with his mother waiting for a bus on Westmoreland Street when he saw the blaze take hold of the building. As he and his mother stood with the other gathering hordes, he remembers thinking that he was watching the end of *The Irish Times*. The following say, the young Kieran was amazed that *The Times* had managed to publish a paper. So impressed was he that he resolved one day to work for a newspaper. He has since worked as Assistant Editor at *The Irish Times*.

Mission of Mercy

The Merryweather Hatfield Fire Engine at Dundalk fire station is beautifully preserved, with its red enamel hooding, brass tubing and ladder. The cracked leather on the old seat is still original as is the leather fan belt inside its impeccably maintained engine which is still in fine working order. The engine itself has a very important place in the history not just of Dundalk Fire Brigade but also Ireland's history.

On the morning of 16 April 1941, firefighters from Dundalk and Drogheda joined their Dublin counterparts in crossing the border to help fight the flames in Belfast caused by one of the most savage air assaults of the Second World War. In total, over seventy firefighters crossed the border on the mission of mercy that spring morning, and whilst the contribution of the tiny brigades in Dundalk and Drogheda has naturally been overshadowed by that of Dublin Fire Brigade in the intervening years, 'It's something which is inculcated in the lore of the brigade here,' says Jim Kerley of Dundalk Fire Service. 'That morning, we sent the only fire appliance we had to help the people of Belfast.' (It was not the first time that the town had sent its firefighting equipment into a war. During the First World War, the horse which had been used to pull the town's firefighting pump was seconded to the war effort and was never returned.)

The handwritten log which confirmed the order to send Irish firefighters into the maelstrom of a Second World War blitz is amazingly neat and concise. Writing at 6.45 am on the morning of 16 April 1941, Mr D. Lennon, the superintendent of nearby Drogheda Fire Brigade noted:

> 'Telephone call from Major Comerford, Chief Superintendent Dublin Fire Brigade, to send firefighting appliance to Belfast to assist at fighting fires, the result of air raids of incendiary and high explosive bombs. He informed me that the Taoiseach (Eamon de Valera) had given written permission for any available appliance to be sent to Belfast.'

By the time the brigades from the South were making their way across the border, Belfast was ablaze. An armada of 180 German warplanes had devastated the city that night, blitzing it with hundreds of high explosive bombs and parachute mines. This was the second time that Hitler's Luftwaffe had turned its attentions to the Northern capital. Eight days before, on 7 April, a smaller attack which left thirteen dead had been carried out, designed to probe the city's defences. The Luftwaffe must have been very pleased with what they saw: a major Allied armament-production city which lay virtually defenceless.

John MacDermott, Minister for Public Security, had noted on his appointment in 1940 that the city was spectacularly unprepared for an aerial assault. Its firefighting equipment had been recalled to London, as voices in Westminster deemed it would be wasted in Belfast. When MacDermott then tried to resupply the city, equipment was severely restricted and in addition there was no time to adequately train the

(*Left to right*) Former Roscommon ACFO Frank Cafferty, CFO Cathal McConn and former Ballaghadereen S/O Fergus Frain.

(Picture: Stephen De Paor)

A memorial to Timmie Horgan (1954–1994) being unveiled in Kilbarrack Fire Station in 2004.

(Picture: Ron McGarry)

Glamorous? A Carlow firefighter shows the reality of fire and rescue work.
(Picture: Paul Curran, Carlow Fire Service)

RTA callouts place an increasing burden on all fire services, particularly on the retained brigades.

(Picture: Paul Curran, Carlow Fire Service)

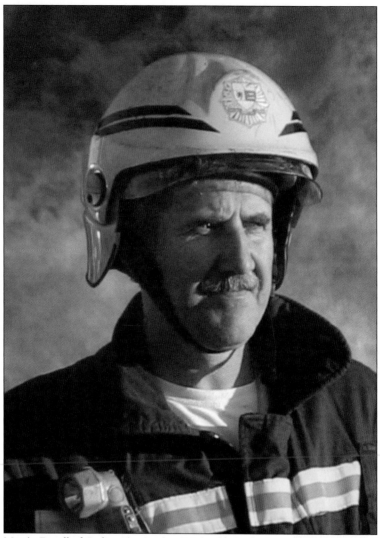

Martin Farrell of Carlow Fire Service. Retained brigades have a thinning cadre of experienced men, and there is pressure to change the existing retirement age from 55 as it is now. This would allow a greater experience level to be passed on to younger firefighters.

(Picture: Paul Cullen, Carlow Fire Service)

Retained firefighter Martin Mulhall. Retained firefighters such as Martin are motivated primarily by a desire to make a difference within their communities.

(Picture: Paul Cullen, Carlow Fire Service)

Retained firefighters tackling a blaze. The retained brigade has broadly achieved equality with full-time firefighters in terms of the standard of equipment and training nationwide.

(Picture: Paul Curran, Carlow Fire Service)

Retained firefighters tackling a truck blaze. Vehicle fires are an increasingly common part of firefighters' lives in the retained brigades.

(Picture: Paul Curran, Carlow Fire Service)

The Irish Times office on D'Olier Street, ablaze, 1951.

(Courtesy of *Firecall* Magazine)

Brian Dempsey (RIP), who was at the forefront of SRT development within the Brigade.

(Picture: Ron McGarry. Courtesy of *Firecall* Magazine)

Auxiliary Fire Service in its use. A concerted programme of shelter construction was begun but by the night of the blitz it still had only half the shelters it should have had. Militarily, there were only twenty anti-aircraft batteries, no fighter squadron, no balloon barrage, no smoke screens and no searchlights. In an ominous letter to the Prime Minister of Northern Ireland, John Andrews, on 24 March 1941, McDermott said, 'The enemy could not easily reach Belfast in force except during a period of moonlight. The period of the next moon from, say, 7–16 April may well bring our turn.'

The ferocity of the assault on 16 April reinforced the notion that the targeting of civilian areas was now an accepted tactic of war. Fatality figures for that night range from 745 to almost 1,000. Outside of London, no other Allied city lost so many of its citizens in a single German attack. As the bombs hammered down on heavily populated areas of the city, fires and explosions illuminated the city for the circling Luftwaffe who patrolled the skies with predatory ease. As the night went on and the rain of death from above continued, the city of Belfast began to feel alone and very desperate. Water mains were destroyed, buildings collapsed, fire met fire to create huge conflagrations. Belatedly, a destroyer was dispatched from Liverpool with fire pumps and much needed full-time firefighters but they would not arrive in time to help control the carnage.

Belfast's battered firefighting service could not cope. The city had a little over 200 full-time firefighters to cope with the devastation with assistance called in from every other area of the province. Shortly after 4.30 am, a rail telegram from McDermott was delivered to the hands of PJ Hernon, Dublin City Manager, seeking the assistance of firefighters and firefighting appliances from the Republic. In a little over an hour, An Taoiseach Eamon de Valera had approved the request and volunteer firefighters from around the city of Dublin were gathering in Tara Street to be addressed by Chief Fire Officer Major Comerford. Thomas Kenny, Dundalk's fire brigade Superintendent, was then called by Comerford and de Valera's directive — 'to give any assistance possible' — was put into action. Dundalk's archives recall the deployment: 'The large trailer pump and five regular firemen left at 7.30 am in charge of the Superintendent. At 11 am the Fire Engine with two regular firemen and four auxiliaries left in charge of the town surveyor. At 11.30 am, ten auxiliary firemen in charge of the station officer A. Halliday left by car.' From the capital, in addition to Dublin Fire Brigade, Dun Laoghaire Brigade also headed north. Five firefighters and one officer volunteered for the long journey.

For the firefighters from the city and Dundalk and Drogheda, conditions on the road north were harsh. The only protection afforded was a windscreen for the driver and, although the air raid itself had finished around 5 am, Belfast was still a city in flames when the crews from the Republic arrived during the course of the morning. Any sort of organisation within the city's emergency services had collapsed and the Republic's firefighters were deployed into a chaotic atmosphere. Edward Lennon, from Dun Laoghaire, later told the Dun Laoghaire Historical Society of the harrowing scenes he witnessed when entering the city: 'I'll tell you what was very sad going into Belfast that morning. There were lorries going in and out of the market filled with dead. They were being laid out for identification.'

DFB man Thomas Kenny remembers the fires they faced that day in East Belfast as the worst they had ever seen. His colleague, Jim Dowling, in later recollections spoke of seeing the glow of fire and destruction over Belfast as they crossed the border. Thomas Colman, also of Dublin Fire Brigade remembers the intensity of the day: 'We were belting at it, because the flames had to be out before the blackout the following evening.' Working alongside their Northern counterparts, the firefighters found themselves in a nightmare world of collapsing buildings, delayed detonations from unexploded bombs and unextricated corpses embedded in the smouldering ruins.

In Sean Redmond's book, *Belfast Is Burning, 1941*, the author relates some of the experiences of firefighters that day. John Kelly, a DFB man, recalls 'human bodies and dead animals lying all over the place.' Another firefighter, Paddy Finlay, escaped serious injury or worse when he climbed a Belfast fire engine ladder to remove a hose from a burning building. Just in time, Finlay realised that it was only the ladder he was standing on that was keeping the wall from collapsing. Another Dublin firefighter, Jim Dowling, recalled a fire in a public house on the corner of Chichester Street. When the firecrews happened upon looters in the area, the miscreants actually asked the firefighters if they would mind turning their hoses upon the premises safe in order to cool it down.

Throughout the day, without any regimented food or rest breaks, they fought to save the city of Belfast. As dusk approached though, they had to return home. Should the Germans return again, and if Irish firefighters should be killed by German bombs, the Republic's policy of neutrality could be severely compromised.

The Merryweather Hatfield which returned to Dundalk on the evening of 16 April was populated by beleaguered, hungry and exhausted men, clinging to the side of the red enamel vehicle. Little did they know that less then three weeks later they would again be crossing the border on a mission of mercy.

The attacks of 4 May 1941 were again stunning in their ferocity but this time the city was better prepared. Stronger air defences and more shelters meant that the death toll was far less then the previous raid; another contributing factor was that the targets this time were more economic then civilian. Nevertheless, the call was once again made for assistance from the Republic and the Merryweather Hatfield from Dundalk was on the road north.

The Germans had attacked Belfast in broad daylight on this occasion, arriving over the city around one in the afternoon. Bombs hammered into the city's dockland area, decimating it and creating massive fires which swept through the Abercorn yard, Queen's works, Victoria shipyard and Alexandria works. In total, 205 Luftwaffe planes pummelled Belfast with 95,992 tonnes of tonnes of incendiary and 237 tonnes of high-explosive bombs. The city had scrambled its Hurricane fighter squadron into the skies before withdrawing them. A breakdown in communication following a bomb in the central telephone exchange meant that this fact was not relayed to air defence batteries that had remained silent for fear of hitting their own planes. For two solid hours, the Luftwaffe attacked and Belfast did not respond. As day turned into night and almost 150 fires raged throughout the city, the North once again requested firefighting and rescue assistance from the Republic. At 3.40 am on 5 May, the crews were once again heading across the border.

In the aftermath of the raids, the governments of the North and South resumed their taciturn relationship with one another, but no amount of partisan politics could get in the way of the city of Belfast thanking the Republic's firefighters for rushing to their aid in their darkest hour. A letter was sent to all the brigades from Belfast's Lord Mayor. In the letter to Dundalk FB he expressed his thanks for the 'very willing and prompt aid afforded to Belfast in the circumstances arising out of the heavy air attack made on that city ... and assuring the council that this token of sympathy and friendship is very highly appreciated.'

On a wider scale, the gesture was widely appreciated and reported on by press sympathetic with both Unionists and Nationalists. The *Belfast Telegraph*, a broadly Unionist paper, said: 'The people will remember the magnificent spirit which prompted fire brigades from Eire to rush to the assistance of their comrades of the North.' The *Irish News*, a Nationalist paper, said: 'A word of high praise is due to the unstinting assistance given by our countrymen in the neutral part of this island to this area. Not only have they been prompt in sending their firefighting units; no trouble is too great for the citizens of Eire when it is a question of housing and sheltering refugees. Never was sympathy so manifest; never pity so practised. We in our day of sorrow thank our countrymen from the South.'

Some months later, the horrors of war were to be visited on the Republic's capital when on the night of 31 May three German bombs were dropped on the North Strand area of the city. In wartime terms, the toll of dead may not be considered great, but remembering that Ireland was a neutral country, it is still shocking to read that thirty-four people were killed, over a hundred people injured and 300 homes obliterated or damaged by this senseless act. Seeing the southern capital suffer, in microcosm, similar atrocities, the firefighters in Belfast sent a telegram offering assistance. The Brigade in Dublin was well capable of handling this incident though and replied that the fire was under control but thanked their Northern colleagues for the kind gesture.

There is an almost Hitchcockian twist to the wartime experiences of Paddy Finlay, one of the firefighters who participated in the mercy dash north, who lost several close family members in the bombing of the North Strand. It is featured in the footnotes of Sean Redmond's short book on the burning of Belfast and also in the *Irish Times*. Towards the end of May 1947, Hermann Goertz, a German agent who had been parachuted into Ireland in 1940 was summoned to what was then known as the Aliens Office in Dublin Castle. He was informed that he was being returned to Germany, obviously against his wishes since the entire country was shattered and in terrible turmoil. He suddenly collapsed and a DFB ambulance responded to the call. The ambulance was driven by Finlay, who attended to the stricken agent and took him to Mercer's hospital, where he died very shortly after. Finlay remarked to the *Irish Times* that he had noticed the smell of geraniums in the ambulance and knew then that Goertz had committed suicide by swallowing a cyanide capsule.

Part Four

Terrible Tragedy

'The End Of The World'

Whiddy Island resembles a gentle, undulating hill in the heart of Bantry Bay. Bantry's cemetery lies on a hillside just outside the town, looking down over the waters of what is one of the world's finest deepwater harbours. The cemetery is dominated by a large white memorial, upon which are inscribed the names of the fifty souls who perished in a maritime tragedy of epic proportions in 1979. Below the names, on a plinth where there are always bouquets of flowers, is a ship's bell which bears the name 'Betelgeuse'.

Probably one of the most catastrophic, and dreadfully spectacular, blazes to have occurred within recent history was that at Whiddy Island oil terminal on 8 January 1979. An explosion aboard the French-registered oil-carrying vessel, *Betelgeuse*, and the subsequent fire destroyed a large section of the ship and the offshore jetty, threatening the island's oil terminal, its population, and the town of Bantry itself. For firefighters, it was a battle which could never be won.

When the M.V. *Betelgeuse* left the port of Ras Tanura in the Gulf of Persia on 24 November 1978 she was originally bound for the port of Leixões in Portugal. The ship was owned by Total Compagnie Française de Navigation (TCFN). Aboard the large tanker, registered at 61,776 tonnes, were approximately 115,000 tonnes of Arabian crude oil. The final voyage of the *Betelgeuse* was beset by misfortune, a prelude to her ultimate disaster. Earlier that year one of the technical managers with TCFN, a Mr Peaudeau, visited the vessel in Bahrain and included in his report the following diagnosis of the condition of the ship: 'It seems realistic to expect this vessel to be kept in operation without technical stoppage until the end of 1979 provided that she does not suffer cracks in the shell plating.' Of the crew he wrote, '. . . excellent spirit aboard this vessel to whom the diversity of the voyage does not seem strange.'

Originally, the vessel was due to dock at the port of Sines, near the Portuguese capital of Lisbon, to offload some of her stock but the weather was so bad that she was forced to continue. Further bad luck awaited at her final scheduled port of call, Leixões, when she was prevented from entering the harbour because a ship had sunk at the mouth of the port, blocking all sea traffic. The *Betelgeuse* was then instructed to make her way to the Gulf Oil terminal at Whiddy Island near Bantry in Cork. En route to Ireland, the ship discharged some of her crew at the northern Spanish port of Vigo before making for

Ireland. Her voyage was hampered further by heavy weather in the Bay of Biscay and it was then discovered that her cargo of oil was leaking. A telegram was sent from the ship: 'Starboard gangway torn out as a result of bad weather. Can see traces of fuel on starboard. Plating probably fractured.' The ship's log also noted the crew's concern when they observed a light film of oil on the surface of the sea. When the leakage was reported the ship was due to dock at the French port of Brest for repairs. The following day though, the ship once again made contact with its authorities to report that they would continue onward toward Bantry as the crew had discovered the source of the leak and mended it. This is the first mention of the structural inadequacies of the *Betelgeuse*, which was to contribute heavily to its terrible demise. The ship continued onward to Ireland and berthed at the Gulf Oil terminal just after 8 pm on the evening of 6 January 1979.

The Gulf Oil terminal at Whiddy Island was situated at the south-western edge of the island. The 120-acre site comprised of a tank farm, with twelve crude oil storage tanks, a power house, pump house, fire station and control office. The offshore jetty itself, to which the *Betelgeuse* was moored, was almost 500 metres in length and consisted of island-like structures grouped together. Collectively they were referred to as 'dolphins'. Dolphin 1 was the most westerly; Dolphin 22 was the most easterly and was also the embarkation and disembarkation point for the jetty which could be accessed only by sea. There was no direct physical connection to the mainland.

The offshore jetty to which the *Betelgeuse* was now moored had originally been designed with its own self-contained fire-suppression system. Basically, its fire pumps would be kept under constant water pressure by an automatic pump and should they be opened the main fire pump on the mainland would automatically start as there would be a drop in pressure. However, in 1970, the system was changed and at the time of the *Betelgeuse* disaster the fire pump had to be started in the control room on the mainland. Therefore the crew on the jetty were unable to start their own fire suppression system, relying instead on external assistance.

Onshore, at the terminal itself, there was a fire mains protecting the twelve crude oil tanks and in addition the tanks were protected by a dry-riser foam system. There was also fire protection and suppression equipment for the pump house, power-house and nearby Ascon jetty, which catered for smaller craft. There was also a fully equipped fire engine and a foam pump and tender. There were also BA sets and portable extinguishers. When planning permission was first sought for the terminal by Gulf Oil in 1966, Cork County Council had made it clear that all necessary firefighting equipment must be supplied by the company. In 1974 the Chief Fire Officer of Cork County Council had written to Gulf Oil asking for details of its firefighting capabilities and equipment. A letter of reply, over a month late, had given a detailed response to the CFO's enquiries. The official report into the disaster stated that it was reasonable to assume from this reply that the facilities were adequate. Demonstrations and exercises of the firefighting equipment took place in 1976, '77 and '78. The terminal also had the assistance of local fire brigades in Dunmanway, Bantry and Skibbereen in the event of an emergency. When the *Betelgeuse* docked at Whiddy Island that winter's evening it seemed that the location's firefighting facilities, with the exception of one major flaw on the jetty, were well prepared and of good international standard.

With the ship safely moored off the jetty, the process of discharging its cargo could begin. Before this could be done there were a number of procedures, as listed on Gulf Oil's safety checklist, which had to be completed. The safety checks were carried out by a Pollution Control Officer (PCO), in this case Mr Jeremiah Desmond. Mr Desmond's job was to ensure that the offloading of the crude oil was expedited in a safe and controlled way. While on board the *Betelgeuse* he noticed that a fan in the pump room was not working and he instructed the crew to repair it before the offloading of the crude oil could commence. The safety inspection also required the PCO to inspect the ship's firefighting capabilities, which Mr Desmond did. The *Betelgeuse* had a decent level of firefighting equipment, with seven foam dispensers on the main deck. In addition the ship was supplied with three firefighting suits and twelve BA sets. The officers on board were also properly trained in firefighting and the ship had a regular code of exercises and drills for onboard fire prevention and suppression which included tests of the equipment. Following the inspection by the PCO, the vessel commenced offloading its cargo of crude oil at 11.15 pm on 6 January. The process was completed at approximately 6 pm the following day. Earlier that day, the ship has started taking on ballast in its tanks. The *Betelgeuse* seemed in fine working order with everything proceeding normally.

When the crew of the offshore jetty changed over at 8 pm that evening, there was no sign of any problems. Mr Tom Kingston relieved Jeremiah Desmond as PCO, Denis O'Leary started work as the Plant Protection Officer (PPO) with responsibility for security of the jetty and Captain David Warner went aboard the *Betelgeuse* as pilot for the night. Also working that night as utility men were Charles Brennan, William Shanahan and Cornelius O'Shea. In just over four hours, all would be dead.

It was just approaching 12.30 am on a moonlit night when Mr John O'Connor and his wife Dorothy returned to their house at the Barony in Glengariff, some 5.5 km from Whiddy Island. From their bungalow, they could see clearly across the bay to the island. Mrs O'Connor was making her way to the kitchen to get some supper when she said that she noticed that there was a fishing boat out: she could hear it, the rumbling sound. She commented that it was strange, but not impossible, for a fishing boat to be out so early on a Monday morning.

When her husband heard the same rumbling noise he went to the windows of the lounge and looked across the bay. 'I do believe there is a fire on Whiddy,' he said. His wife went across to the window and looked out, saying that it must only be a twinkling of the light. Her husband had gone to the bedroom for an alternative view and called out, 'It is a fire there.'

Mrs O'Connor opened the French windows of her lounge and went into the garden. She realised that there was indeed a fire, but it was quite small and as she said in the subsequent report it seemed of 'manageable proportions'. Then, as she was looking at the fire, a huge roar echoed across the bay and the fire raced across the tanker which was visible clearly beneath the moon. She ran back into the house to raise the alarm.

Back at Whiddy Island, all hell had broken loose. There are a number of theories as to what started the blaze. After the disaster, the truth was clouded by claim and counter-claim with no clear evidence pointing to once tangible event which triggered the calamity. The official report into the event judged that the initiating occurrence was the 'buckling of the ship's structure at or about deck level and in the way of the permanent ballast tanks forward of the ship's manifold. This was immediately followed by explosions in the permanent ballast tanks and the breaking of the ship's back.' From evidence gathered by eye-witnesses, it was just after 12.30 am when disaster struck the *Betelgeuse*. A fire, at first very small and localised, was observed close to the ship's manifold, at the open ends of the pipelines. The flames gradually grew in scale and intensity along the starboard side of the ship, accompanied by a large plume of dense black smoke. This acceleration of the fire was what Mrs O'Connor saw from her garden. The report into the tragedy accumulated multiple eye witness statements, but that of the O'Connor's was the earliest reported notice of the fire.

It was shortly after 1 am that a massive explosion occurred, followed by a number of subsequent smaller explosions. Eye witnesses referred to the initial large explosion as thunderous. Mr Frits van Os, who lived over 6 km from the jetty, said that his house shuddered with the force of the blast. The fire services' first involvement that night was at about 12.45. Patrick Keane, a member of Bantry Fire Brigade, noticed the blaze and having run to the garda station to alert local police he then ran onwards to the house of Hugh McCarthy, Bantry's fire officer. They then drove down to Bantry pier and, seeing the scale of the inferno, made their way back to the garda station where they called James Hodnett, Bantry Fire Brigade s/o. Mr Hodnett arranged for the siren to be sounded and the town's firefighters began to gather. At the garda station, Mr McCarthy called the brigades in both Dunmanway and Skibbereen, seeking assistance. He then collected equipment from the station and made his way to the pier where his personnel had mustered. At the station he had met another firefighter, Thomas Muckley, and the two men were gathering the station's gear together when they heard the massive explosion which broke the back of the *Betelgeuse*.

The biggest problem for the local firefighters that night was not preparation for the battle ahead, but marine transport to get them out to the jetty. Some firefighters did not arrive on scene until well after 2 am. However, their actions on the night were credited as being as efficient and forthright as possible. The report into the incident relates that no attempt was made by the dispatcher at Whiddy Island terminal or by anybody on the Gulf Oil jetty to contact the fire services.

As is detailed in the inquiry into the incident the exact events surrounding the triggering of the disaster are shrouded in uncertainty. The official report into the disaster pointed to the buckling of ballast tanks at the port and starboard of the ship as the event which triggered the ship's disaster. The last moments of those onboard the *Betelgeuse* and the offshore jetty are unclear. However, the fact that the coroner's verdict on the bodies that were recovered indicated death by drowning rather then incineration suggests that there was a brief but ultimately futile chance for those onboard to escape.

One of the local men who died, the ship's pilot David Warner, had been onboard the *Betelgeuse* when the fire took hold and was one of the first identified in the aftermath. Moments before the explosion destroyed the vessel, his transmission was picked up by a nearby vessel, the *Bilbao*. Warner's voice was heard: 'Get off, for God's sake, get off.' Five other local men died together in the control hut on Dolphin 22, which bore the brunt of the *Betelgeuse* explosion. The disputed testimony of the terminal's control room operator, John Connolly, recalls eight men being enveloped in flame aboard the jetty. The firefighting capabilities on the jetty were never activated. To do so, the fire pump at the oil terminal would have had to be switched on, so the pumps could be pressurised. This was never done.

The fire is believed to have started forward of the manifold on the ship before it spread rapidly prior to the major explosion which buckled the vessel. Speaking to the press soon after the disaster the president of TCFN, Louis Bouzol, confirmed that the ship did not have the 'inert' gas system regularly used in tankers to prevent the explosion of the combustible gases which can accumulate as oil is discharged. The tribunal of inquiry pointed to incorrect ballasting procedures on the night as another contributing factor.

During the initial fire, caused by structural failure and explosions in the permanent ballast tanks, the ship had already broken its back and settled into the water. The flames at this stage emanated from oil which had collected in the water around the centre of the ship. Firefighting capabilities aboard the ship were severely disabled as the structural collapse had fractured the fire lines while the thick black smoke and water meant that the jetty crew were unable to successfully challenge the fire from the jetty. The flames were initially quite low and did not threaten the jetty, but between 00.50 and 00.55 the flames roared toward the jetty platform, including the crew of Dolphin 22, and consumed it in flames that reached as high as sixty-one metres. The massive explosion caused a huge release of oil which bore the flames dangerously close to the Whiddy shore.

The only vessels which were in a position to mount any kind of rescue mission were the tugships *Donemark* and *Snave* at the Ascon jetty and the *Bantry Bay* which was moored nearby. These vessels were marine workhorses, without any firefighting and fire protection equipment. Nevertheless, just before 1 am they departed the Ascon jetty for the *Betelgeuse* which was by now engulfed in swathes of flame. The *Bantry Bay* managed to get close to the bow of the *Betelgeuse* and near to Dolphin 1, the point on the jetty which was closest to land. Joseph Kelleher, skipper of the *Bantry Bay*, described flames shooting 'hundreds of feet into the air'. From what he could see, there were no signs of life onboard the ship. With the flames now spreading rapidly, the small tugboats found themselves in an increasingly perilous environment. There was no way to reach Dolphin 22 and the doomed jetty crew. Then a small explosion echoed from the *Betelgeuse*, and the crew of the *Bantry Bay* decided it was time to move out. Other boats soon came close to the fire, searching the sea for survivors or victims. The efforts of the tugships in venturing so close to the inferno were commended for their courage. However, there was little or nothing that could be done when they arrived. At that stage, the ship had suffered a cataclysmic explosion and the dead had already been claimed.

Noreen O'Leary's husband and son, nearby residents on Whiddy, were caught up in the impact of the massive explosion on the *Betelgeuse* but survived relatively unscathed. The pictures in the papers of the following days depicted destruction on a massive scale. The O'Leary family were the closest residents to the Whiddy Island terminal, and they described an almost apocalyptic scenario to reporters:

'I looked out the window and saw huge balls of flame dropping right into the farmyard. I was looking out onto the jetty from my bedroom and I thought the world was about to come to an end.' When Noreen's husband and son returned they told the family to grab what they could and leave. 'We ran out, leaving the house and the cattle, leaving everything and we got to the shore as quickly as possible.' The family made their escape in their own boat. Two elderly residents of the island, aged 94 and 86, were among the last to reach safety as smoke and falling flame threatened the entire island. All sixty-one residents of the island were evacuated on the night of the disaster.

As the islanders were leaving, the firefighters and other emergency personnel were just arriving via a small flotilla of local boats. They had been hampered by a lack of transportation from the pier in Bantry and again when they got to the Ascon jetty on Whiddy as the vehicles at Gulf's fire station were not in working order. When the crews finally arrived at the terminal they began cooling the oil tanks on the tank farm.

The fire had already progressed to a phase which would make any firefighting measures totally futile. The intense heat had warped girders of solid steel on the nearby jetty and drops of molten fire mingled with the burning sea around the tanker. The ship itself had collapsed into the inferno, split in the middle, its bow and stern jutted into the sky, shrouded in flame and black smoke. At this stage nothing could be done for those on the *Betelgeuse* or on the jetty. The tribunal into the incident corroborated this, stating, 'There was nothing the brigades could have done to minimise the disaster had transportation been available.'

The blaze aboard the *Betelgeuse* raged through the night and into the next day. The first images of the scene in daylight were dominated by the pillar of black smoke reaching skyward. The fire itself burnt out around 8 pm that evening. For police and firefighters on the ground, it was the start of a grim day. The tragedy would only leave behind twenty-seven bodies from the fifty that it claimed. Most of the ship's crew, according to the post mortem, died from drowning while the jetty crew perished in the flames. On the day following the disaster, thirteen bodies were recovered, seven on the north shore of the island and six near to the jetty. On 9 January, a search party made its way to the mangled wreckage of Dolphin 22. Another eight victims were found here, five together inside the security hut.

For those with loved ones working on the offshore jetty that night the torture was just beginning. The wife of the late James O'Sullivan, Jenny, was going to bed when the explosion struck. Talking to the *Irish Independent*, the distraught woman said, 'I looked out and saw massive flames and I immediately thought of Jimmy. Then I ran down to my brother-in-law's house and he drove me to the pier, where I stood all night looking out to sea. I tried not to give up hope. But then when I heard people talking about all the dead, I knew the worst had happened.' The pair had only been married for two weeks.

Terrible sadness and poignancy also surrounds the story of the

Shanahan family. Liam Shanahan was not even supposed to be working the night he died. His wife was heavily pregnant and expecting any day. However, a colleague was suffering badly with a toothache and asked Liam to cover for him. When his wife dropped him off at the terminal that evening, she started to suffer from contractions and checked into Bantry hospital, leaving word that Liam be told when he was finished the night's work. That night, as tragedy claimed the life of her husband, Teresa Shanahan gave birth to a baby boy. Mrs Shanahan was only informed of her husband's death on Tuesday, two days after her husband had died. 'She is aware that there has been an accident in which her husband was involved, but it will be today before she is told the final truth,' a hospital spokesperson told the press. Marianne O'Leary was another with her life shattered by the tragedy. She had been engaged to be married to terminal worker Charles Brennan when he died aboard the jetty.

Due in no small part to its remote location and problems of accessibility, firefighters, when they arrived on Whiddy Island, were simply embarking on a salvage, recovery and damage limitation exercise. However, less then ten years earlier, on the morning of 6 November 1974 the country's largest oil and petrol stores outside Whiddy Island in Dublin's North Wall, were also subjected to a terrifying inferno when the Calor Kosangas plant on the North Wall literally exploded. Triggered by a fracture in the gas-supply line the subsequent explosion catapulted red-hot gas cylinders into the air as firefighters raced to the scene. Incredibly, there were no deaths as a result of the incident and few injuries. According to the *Irish Independent* of the following day: 'the first explosion sent sheets of orange flame hundreds of feet into the air. The massive blast was heard as far away as Dun Laoghaire. The entire dockland area was evacuated minutes after the first blast and, as explosions rocked the city at the rate of one a minute, hundreds of workers and their families were moved out. Twelve units and fifty firemen from Dublin fire stations hit the inferno with thousands of gallons of foam and water.'

The firefighters trained many of their hoses on the 25,000-gallon gas tankers, which were only as far as twenty yards from the exploding cylinders. If one of these tankers was to explode, in an environment such as a densely populated city centre, the result could well be cataclysmic. As the flames raged out of control for several hours, gardaí insisted on evacuating people from the area and implemented a half-mile exclusion zone around the fire. The initial fear, before the cause of the explosion was known, was that the city had been bombed, as memories were still raw from the bombs which had killed so many in Dublin and Monaghan the previous May.

It was the swift actions of the Calor Kosangas staff, in addition to the sharp tactics of the firefighters, that brought the blaze under control before 4 am. However, the whole incident further shook the fragile confidence of the city. The *Irish Independent* was in no mood to lighten people's spirits when on 7 November it published a story with the headline, 'Is Dublin Sitting on a Time Bomb?' Setting the tabloid tone, the article's writer, Jim Farrelly, went for the jugular: 'Is Dublin city sitting on a 40-million-gallon petrol bomb? Could almost a million greater-Dublin dwellers be annihilated because of a simple gas leakage or oil spillage in the petrol storage zone at Dublin's North Wall?' he asked. Apparently not, according to the military, petroleum and port managers who were interviewed for the article. Mr John Donovan of Esso said that the city was sitting on no potential bomb: 'The whole layout of the depots of the oil companies is in accord with international standards for the control of petrol fires.' An army ordnance disposal expert was adamant that a fire at one depot or storage tank would not trigger a deadly chain reaction of fires and while the exact source of the spark that ignited the leaking gas was a mystery subsequent to the fire, one source was quickly discounted: hobnail boots. 'Hobnail boots are forbidden at Kosangas, lest they make sparks,' said Mr Joseph Murphy, the Dublin Port and Docks Board General Manager. It appears the explosion and fire that night was more spectacularly destructive then deadly dangerous.

Stardust

'I remember thinking, this is it, I will never get out of here alive, never … I had completely lost my sense of direction and I hadn't a clue where I was going. I couldn't scream because I couldn't get enough air in to scream. At this stage, I was beginning to feel like I was going to pass out. It was becoming nearly impossible to breathe. I couldn't see at all. My eyes just wouldn't open … Everyone around me was screaming and panicking and I remember thinking, why can't I scream? I thought of my parents and how they would feel but the worse was thinking I am going

to die in this horrible place and there isn't a damn thing I can do about it. What on earth had I done to deserve this from God?'
(Excerpt from a statement by a Stardust survivor to the Stardust Victims' Compensation Tribunal).

When asked what effect the Stardust fire had on him, a now-retired DFB officer, with over forty years' experience, said, 'It was at a fire on Dorset Street when it all became too much for me. I just couldn't handle any more death. It wasn't the fire I was attending, although there were fatalities, it was just that I had seen too much death and it had taken its toll on me. It was a delayed reaction to what I had seen just a few days earlier. It was a delayed reaction to Stardust.' At the end of our conversation he said, 'You know, I have probably just told you more about Stardust in the last few minutes then I have ever told anyone in my crew. Do you know why that is? Because I don't want to talk about it; I remember it clearly, but I find it difficult to talk about things like that.'

Today's recollection of one firefighter who was there still provides only a small snapshots of the horror of 14 February 1981. Stardust, one of the most tragic fires since the death of 35 children in a Cavan orphanage in 1943. So much material has been written, recorded and said about Stardust that it verges on the exhaustive. The events on Kilmore Road in Artane have been the subject of two tribunals, vast tracts of newsprint, books, poems, songs and an RTÉ drama series. The hurt over Stardust is still raw, the death of forty-eight people and the scores of life-shattering injuries never forgotten.

However, it still retains a grip on the memory of the nation and the life of any firefighter in Dublin Fire Brigade. 'There can be nothing written about this brigade without mention of it,' says a current crewmember with only five years' experience. 'It's a cultural thing within the brigade, because after Stardust the organisation underwent massive change.'

The Stardust club's disco and live entertainment nights were one of the few sources of entertainment in this working-class area, during a time when jobs were scarce. The Stardust was part of a larger complex which had been converted from a factory. Valentine's night saw the venue packed to capacity, but due to the nature of its construction and the materials used on the interior, it was a death-trap. The tribunal into the tragedy said that the fire was almost certainly caused deliberately, one of the seats being ripped open and the flame-friendly synthetic materials set alight. Two minutes later, the fake ceiling of the club was burning and falling upon panicked revellers as they sought to escape.

In DFB HQ control, three firefighters and one sub-officer were dealing with the night's workload. FF Mooney, who answered the call about the fire, said in his account that the phone rang at 1.43 am. 'I said, "Dublin Fire Brigade". This chap rang in and he said, "there is a fire in Stardust." I said, "where is the Stardust?" He said, "it is in Artane." I said, "what is on fire?" He said, "the whole place is on fire."'

Following the call, HQ Sub-officer Hughes called Kilbarrack and North Strand stations and told them to respond to a fire at the Stardust club on Kilmore Road. The Kilbarrack appliance took between five and eight minutes to reach the Stardust, which was by now venting flames. Kilbarrack s/o Mooney said that he saw many people in a hysterical condition on the road, many of whom were suffering from burns. He radioed HQ and put in a call for assistance from throughout the district.

The scene awaiting firefighters approaching the Stardust was chaotic. The security barrier in front of the building was still in place and people pushing forward moved the barrier towards the front of the appliance which obstructed it. When the firefighters alighted from the vehicles and started to make down hoses to tackle the fire, they were mobbed by people asking them to find their friends, family or loved ones. The flames were now shooting from the sides of the building and fire crews trained their hoses on them to try and drive them back from the doorways so they could effect an entry. Arriving ambulances were simply unable to cope with the number of casualties and the prevailing mood of panic. Any public or private vehicle became an ambulance that night. The intense heat at the exits of the building made it impossible for firefighters to enter until they had been cooled. They then learned that there were people still alive in the toilets of the premises. Firefighter Noel Hosback, still with the brigade, succeeded in navigating through the burnt interior as far as the men's toilets where he allowed people to access fresh air through his breathing set so they gain enough strength to get out of the building. He led them through to the spirit store where they were helped out the window.

Other firefighters who entered the building came across the first of many bodies. 'I could describe to you the condition of some of the people who we found that night, but what good would that do,' says a retired DFB district officer, who was a firefighter on scene. 'We found them in all sorts of conditions. It was terrible. I remember two in particular; we found them quite close to the door. Their bodies had been badly burnt but you could see that they died embracing each other.' Stories abound of the horror in the Stardust, but firefighters are reluctant to elaborate on specific, harrowing details. 'It was chaos, both inside and outside,' he says. 'These were young people on a night out and when that place went up like it did, many of them never stood a chance.'

Not on duty that night for Dublin Fire Brigade was Jim McDermott. Three of his children had gone to the Stardust for their Valentine's night out. The three bodies of Willie, George and Marcella were recovered by firefighters. Their father retired from DFB on medical grounds not long after and has since passed on.

The aftermath of the Stardust was a time of grief followed by painful reflection and introspection. The Tribunal into the tragedy was set up the following day. Over the subsequent weeks and months the full story of negligence and incompetence, compounded most likely by arson, emerged. It transpired that the owner of the Stardust, Eamonn Butterly, had no idea of the Corporation by-law which stipulated that he needed to assign staff to supervise people in the event of fire. He also admitted that he had absolutely no idea of a Department of Local Government directive on fire protection standards and that there had been absolutely nobody in the Stardust to supervise fire safety procedures. Counsel acting for victims' families stated that at least one of the doors in the club had been locked and three others were locked or difficult to open, particularly in a crowded environment in which the lights had failed. Two of the exits were also obstructed by skips. Fire experts who carried out tests on the cause and spread of the fire confirmed that the carpet tiles used on the walls of the Stardust were highly flammable and responsible for the rapid development of the blaze. In addition, windows in the premises were in many cases covered with metal or were barred. While the negligence of the owner is shocking, his poor under-standing and implementation of fire safety law was commonplace, such was the lack of adequate fire prevention in the city at that time.

In this and other aspects Dublin Fire Brigade was heavily criticised, though mostly via the bad management of Dublin Corporation. 'The members of the fire brigade carried out their duties on the night of the fire in a dedicated manner and, in the case of those firemen and fire officers who took part in the actual evacuation of the people from the building, at considerable risk to themselves . . .' the report of the Tribunal stated. But, in the same paragraph, the report went on to say: 'The evidence clearly established, however, that the management (including the general organisation), the training and the equipment of the Dublin Fire Brigade had not been modernised so as to enable it to cope with a disaster of these proportions in a satisfactory manner.' The report criticised a lack of training, equipment, a cohesive command and control structure and an out-of-date control room. The most damning paragraph said: 'The Tribunal has, accordingly, come to the conclusions that, although injury and discomfort to a number of survivors would probably have been avoided by a more efficiently mounted rescue operation, no avoidable deaths were shown to have resulted. It must be emphasised, however, that had the course of events been different the consequences in terms of human lives lost as a result of the deficiencies in the DFB revealed at the Tribunal's hearings might have been calamitous indeed.'

Firefighters who did their best to save who they could that night are reluctant to talk in terms of their own personal cost. Stardust stays with every firefighter who was there to this day. 'I remember one man coming out, he was literally burning, skin peeling. He took off his clothes, we gave him a BA set and we hosed him down, thinking back to some of the things which I saw inside that club, he was one of the lucky ones' says one DFB officer. 'I also remember the mood back in the station, lads in shock, trying to take it all in,' the former D/o adds. 'I remember the hypocrisy in the week afterwards too. We did our best that night, whatever people say. Later on the following week we got a visit from a Fine Gael TD who praised us for the great work we did, to congratulate us on a job well done. Firefighters aren't very comfortable with that sort of thing. A few months later I heard him on the radio saying that we were all getting paid too much, that's the sort of hypocrisy I'm talking about.'

In the wake of such findings, there was a need for a complete overhaul of Irish fire safety legislation. It is widely understood that the Fire Services Act of 1982 was published as a direct response to the Stardust tragedy. This is not entirely true: while Stardust was a major catalyst in moving the act forward, the genesis of the whole process was the Noyeks fire, almost ten years previously. If there is anything remotely positive to be gleaned from the horrific events at Stardust, it is that serious fire safety legislation now exists for entertainment venues. A fire safety code must be incorporated into all regulations governing the construction, design and management of entertainment venues. However, such legislation always needs to change with the times and the current legislation is now well over twenty years old, a fact highlighted by fire officers as far back as their annual conferences in 1999 and before.

For Antoinette Keegan, who was eighteen when she lost two sisters at Stardust, the fear is still there that a similar tragedy could occur again. 'The first thing I do when I visit a pub or a club is to check all the exits. I still tense up if I find myself on a dance floor.'

Noyeks: 27 March 1972

The picture of Dublin firefighters recovering the body of a victim from the tragic blaze at Noyeks timber showroom on Parnell Street on the afternoon of 27 March 1972 seems to hail from a bygone age. The firefighters are clad in little more then sou'westers, and wear helmets which, in appearance, recall much earlier times. But this was a terrible day. Before the Stardust fire in 1980, this was the worst recorded loss of life in a fire in the capital city. Eight people perished in a fiery inferno, seven of whom were young women. Many of those saved that day were thankful to members of the public who reacted quickly to a blisteringly intense and fast-burning fire. Indeed, it was the DFB who were criticised, albeit mostly unjustly, in the aftermath of the tragedy.

In reality, Noyeks was a disaster waiting to happen. This is corroborated by a report from Dublin's Chief Fire Officer, Thomas P. O'Brien, which he submitted in August 1972. The building consisted of three floors and a basement and was constructed from materials which were common at the time, although perilously conducive to fire, with timber floors and an asbestos and timber roof. The building also had central heating but it was out of order and in its place a number of gas fires had been placed on different floors, so there were live flames on the premises. To compound the dangers, quite large quantities of flammable products were being stored on the premises, including petroleum products, but it did not have and had not applied for, a petroleum storage licence. It is understood that the blaze started when a tin of Bostic adhesive, highly flammable, fell from a shelf near one of the fires. Like many items in the premises, this was stored in a highly dangerous fashion: instructions on the adhesive warn that it is not to be stored near flame.

At 14.51, a call to Dublin Fire Brigade to respond to the fire was logged and two pumps and a turntable ladder responded immediately, Dorset Street being the closest station. Six minutes later, due to the scale of the incident, a district call was sent out and another appliance, one emergency tender and two ambulances were dispatched.

So intense was the heat from this terrible fire that it cracked windows in premises on the opposite side of the street. When the fire started, following the adhesive spill which was then ignited by a gas heater, it spread with frightening ease and speed trapping those on the upper two floors. Between forty and fifty people were working in the building when the blaze started.

Staff at Peat's electronics, across the road from where Noyeks stood, were amongst the first to become aware of the fire. Harry Peat, the firm's director said he noticed smoke billowing from the building before flames exploded into the air. At this stage he could see people trying to escape from the smoke-blackened windows on the upper floor. Along with his brother Geoff and other employees, they rushed to the building with ladders to assist the frantic escape attempts. His wife, Isabel, saw two girls screaming for help as the flames closed in: 'They were trying to open the windows and were clawing at the Venetian blinds. I saw them there for about ten seconds and then there was a huge roar and the floors collapsed; they hadn't a chance.' With the help of Peat's staff, eight or nine people were brought to safety. One was the company director, Ivor Noyek, whose body was found in front of his blazing premises; he was injured but alive and was dragged clear of the flames and searing heat. His brother, Sam Noyek, and the sales manager, Joseph Nagle, were also fortunate. Both men managed a narrow escape by edging past the flames as they raced down the stairs which were burning beneath their feet. Geoff Peat remembers seeing a man dash into the inferno to rescue another, burning his hair off in the attempt.

The first DFB officer on scene was s/o Michael Gibney from Dorset Street. When his appliance arrived, the entire façade of Noyeks was ablaze and the flames were shooting halfway across Parnell Street. So intense was the heat that he had to keep to the opposite side of the street. It was noticed that the fire was now threatening buildings to each side of Noyeks and that even buildings across the street were showing the effects of the heat. He instructed his firefighters to set up two hose lines and to train one of them on a neighbouring building which was beginning to smoulder. He was then told by a frantic member of the public that people were jumping from the building onto King's Inns Street. Going there immediately, Gibney saw one young woman lying on the ground being attended to. The entire front of Noyeks facing onto that street was also ablaze, yet he did not notice anyone screaming for rescue. It was at this stage that he sent out the district call, just six minutes after the initial call had been received. The building was now completely engulfed in fire on two sides and at no time did s/o Gibney consider it at all safe or prudent to send his firefighters into that building.

Two other appliances from Tara Street, part of the initial dispatch, arrived under the command of D/o Michael Conroy who was now in charge of the fireground. In his statement to the CFO for his report, Conroy repeated Gibney's statement that the building was gutted with fire at this stage, he did not see anyone at the upstairs windows of Noyeks and he has no doubt that the eight people who perished that day were dead before his arrival. He added that he was not delayed in getting to the scene by traffic and that there was no problem with water pressure, which he described as excellent. The impression from these firefighters' statements is that everything was done in the swiftest possible fashion by Dublin Fire Brigade; they were quick, but the flames were quicker.

When Conroy hit the fireground, he told his men to get to work with further jets hitting the fire, including one from a turntable ladder on the Parnell Street and King's Inns Street sides of the building. There were now seven water lines trained on the blaze.

Another threat became apparent as the flames began to threaten nearby tanks containing thousands of gallons of oil and dozens of gas cylinders. Two more DFB officers had by now arrived on scene, Second Officer Larkin and Third Officer Kiernan. It was just after 3.30 that the Brigade put out the call that the fire was on the retreat: 'Fire under control, at least five people unaccounted for,' is how it's phrased in the CFO's report. At this stage the roof of Noyeks was gone and the two upper floors had collapsed down upon the ground floor. Parts of the ground floor to the rear were badly burned. The brigade requested the attendance of Corporation's Dangerous Buildings section for an evaluation.

At this stage, those who had escaped from the fire had been taken away in waiting ambulances or were receiving treatment on scene. For the firefighters, the gruesome task of recovering the victims remained and from the condition of the fireground it was clear it would only be a recovery exercise. Shortly after 4 pm, when the fire had been brought under control, firefighters began to enter the building. At this stage a huge crowd had gathered, including employees and friends of those who didn't make it out. Gardaí had to use loudspeakers to keep the crowd back and give firefighters space to work. At around 4.30 pm the first victim of the fire was recovered. A DFB officer on scene said that many of the victims at the Noyeks fire were too badly burned to be initially recognised. Inside the smouldering ruins of the firm, firefighters used rakes to sift through the charred and smoking debris, intermittently stopping work to recover and remove the remains of another victim. The remains of the eighth victim were recovered by 5.30 pm. The search continued until 11 am on 29 March to ensure that all remains had been recovered.

The devastation at the scene was quite incredible. Pictures in the *Irish Independent* of the following day show a huge, smoking shell of a building, the plastic lettering at its side charred and blackened. Other pictures show blanket-laden bodies loaded into the back of ambulances. The eight victims of the fire were all found in close proximity to each other, they evidently all perished when the top floor of the building collapsed. They were: Mary Culleton, Mary Leader, Martha de Burgh Whyte, Patricia Gore, Marie McNally, Dolores Finnerty and Ann Condon. The only man to die in the fire was James Pryal.

In the aftermath of the tragedy, the recrimination and inquiries started. Relatives of the dead and injured and elected officials called for an official and full-scale inquiry to be launched. There were initial fears that the fire could have been a result of arson but this was discounted by the gardaí. A special report into the incident was commissioned by the City Manager, Matthew Macken, and it was made available to the Minister for Local Government, who was responsible for the fire services. Frank Cluskey TD said that an inquiry into the fire at Noyeks should be launched as there were other buildings in the city which were constructed in a similar fashion and another, similar, tragedy could occur again. The DFB, as is detailed in O'Brien's report, was in a lose-lose situation. Firefighters had responded to the fire as quickly as possible, calling for more assistance within six minutes from the fireground. They had also achieved the not-unimpressive feat of effectively knocking down a major blaze in a three-storey building in less than 45 minutes. As the CFO says: 'I am satisfied that there was no delay in the response of any of the fire brigade appliances. I am of the opinion that the first pump (appliance) arrived within a very few minutes of the fire call being received and that this was immediately followed by the other two appliances forming the first attendance. I am satisfied that no unusual traffic difficulties were encountered. I am of the opinion that when the first pump arrived the building was so well alight that no rescue was possible. I am satisfied that the fire was prevented from spreading and quickly surrounded, the 'fire under control' message being returned forty-one minutes after the receipt of the initial call.'

These considerations aside, the fire at Noyeks on Parnell Street nevertheless was a watershed in firefighting in the city in many respects. The following year, 1975, a Report on the Fire Service was issued which used this fire as one of the key driving factors for change. *The Irish Times* of the following day said: 'This fire must inevitably stimulate further discussion about the general question of fire prevention, fire protection and firefighting. All the information to hand suggests that Dublin Fire Brigade and the gardaí acquitted themselves creditably. Evidently the fire spread with remarkable and terrifying rapidity.' The paper went on to reluctantly point the finger of blame at a chronically ignored fire service: 'For many months there has been mounting disquiet about our fire brigades and the Minister for Local Government seems strangely insensitive to it. The men who know the service best, those who operate it, have called urgent attention to its inadequacies and proposed reforms. These reforms have been urged, separately and insistently, by the full-time firemen, the part-time men in the smaller brigades and the second officers. Training has been particularly criticised; firemen are insufficiently trained and there is no formal training here for officers.'

Dublin and Monaghan
On Friday 17 May 1974, mass murder came to the streets of the capital in the shape of car bombs on Parnell Street, Talbot Street and South

Leinster Street, which exploded at the height of rush-hour in the midst of a bus strike, slaughtering twenty-six innocent people. A short time later that evening, terror and death was visited upon the people of Monaghan town in another explosion which murdered seven. The largest explosion in Dublin occurred near to where a monument to the victims stands today, near Guiney's shop on Talbot Street, close to Connolly Station, which killed fourteen people including an unborn child. At Parnell Street, near the Wellcome Inn pub, another large car bomb exploded, killing twelve. At South Leinster Street a bomb inside a sports car detonated, claiming a further two fatalities.

Joe Broughan, D Watch HQ DFB, speaks in a low tone as he remembers the outrage which shattered the city and destroyed scores of lives. 'To be honest, it was one of my very first callouts,' remembers Broughan, who has served in DFB HQ for over thirty years. 'I was only in the brigade a very short time. I remember being in the fire engine, racing across Butt Bridge, when the second bomb detonated.' The front of Liberty Hall seemed to visibly shudder and reverberate from the blast as the stunned firefighters raced onward. 'I do remember fear, definitely, I was scared and no doubt about it,' adds Joe. The scene on Parnell Street was one of complete devastation. 'One of the first sights I remember was a man, his clothes in shreds, running almost naked down the street with shards of glass protruding from his bloodied back. I went about my work in a condition of absolute shock that day, picking up the dead and the wounded. I saw things that day, some of the worst I have ever seen, that are simply indescribable.'

The city had been on alert for some time, aware that the unrest north of the border was bound to be visited on the Republic at some stage. These terrible bombs were not the first: on 26 November 1972, a trip to the cinema near O'Connell Bridge became a scene of terror when a number of people were injured by an explosion at the side door of the building. There were no fatalities that day but six days later, terrorists had claimed their first victims. Just before 8 pm a car bomb detonated at Eden Quay, close to Liberty Hall. Eighteen minutes later, a second bomb went off at Sackville Place, just off O'Connell Street. It was the second explosion that claimed the lives of CIE bus conductors George Bradshaw (30) and Thomas Duff (23) in addition to injuring well over a hundred others.

The DFB had absolutely no training in dealing with incidents like this and had responded as they would to any other callout. (Their counterparts north of the border were learning the hard way that emergency workers and law-enforcement personnel were favoured targets of secondary devices which would detonate shortly after the original, normally when emergency crews were on scene.) In addition to the carnage which the explosions caused, the sheer force of the 1972 blasts threw the city into chaos. Fire and security alarms were triggered throughout the city centre, stationary cars became suspect devices and many major buildings had to be evacuated. While they were not the targets of a second device during the callout, it had been a baptism of fire for the brigade. They had seen the horrors of a terrorist attack first hand.

Terror struck again in January 1973 when seventeen persons were injured and Thomas Douglas, 25, was killed in a car bomb explosion. Terrible as they were, these attacks paled next to the 1974 attacks, which, before the 1998 Omagh bombings, claimed the single largest toll of human life of any day of the conflict.

Like Broughan, Tommy Ellis was coming to the end of another day's work in Kilbarrack when the call came in to respond to an explosion on Parnell Street. 'When the call came through, myself and a former colleague (Denis McGrath RIP) responded in the ambulance. All we knew at that stage was that there had been an explosion on Parnell Street. On the way though, details began to come through on the radio of exactly how bad this was.' Ellis describes a scene of pure chaos, the dead mixed with the living and the barely alive. 'No matter what training is given to deal with traumatic or violent incidents, absolutely nothing can prepare you for something on that scale. One of the first bodies I saw was a man against a car, his head on the road beside him.'

The ambulance from Kilbarrack was one of the first to arrive on scene and although there were plans to deal with major incidents such as this, implementing them in such a hellish environment was very difficult. 'There was a moment when I remember just standing there looking at this horrible scene and not quite knowing what to do or where to start. The training we had then is nothing like the paramedic training which our lads have now. It was far more basic and so was our equipment.' As the emergency response level grew, Ellis remembers a DFB officer telling him to just get the dead off the scene as quickly as possible. 'My mind kind of goes blank after that because I went into some sort of automatic mode, we just had to get this job done,' he says. As he sifted the living from the dead, his mind entered a passively alert mode, common for emergency personnel on the scene of major incidents. 'I picked up bodies and what was left of bodies, that's all I know. Some were whole, many were not.'

As the ambulances transferred the dead to the morgue at Jervis Street hospital and elsewhere in the city, the scale of the attack on the city was becoming clear with reports about the bombs on Talbot Street and South Leinster Street. 'One of our greatest fears at the time was that there would be another bomb while we were working. It was very frightening.' Ellis and McGrath were then dispatched to South Leinster Street to see if any more assistance was needed there. As Tommy drove the ambulance he remembers hearing Denis in the back. 'I remember we had to clean the ambulance, because after what we had just been involved in you can imagine what it would have been like. So as I drove the ambulance, Denis was in the back trying to clean it as best he could, to at least clear away most of the blood.'

From eyewitness descriptions, the only figures moving on a stricken Parnell Street on that terrible afternoon were the dark-clad firefighters, ambulance personnel and gardaí as they moved among the afflicted and the dead. Joe Broughan was on the scene for maybe only half an hour — the shift system at the time meant that another crew was due to come on scene at 6. 'It was a terribly surreal day. I recall returning to the station after doing what we could and after other crews had arrived. Then, I cycled home.' To this day, he doesn't recall that cycle home to his home in Artane, his mind was, as he describes, it 'completely numb'.

Eyewitness accounts from the scene on both Talbot and Parnell Streets are harrowing in the extreme. Newspaper reports in the *Irish Times* and *Irish Press* the next day spoke of streets littered with personal belongings, the dead and the dying. *The Times* described two victims found after being blown through the window of Guiney's clothing store on Talbot Street, the two bodies so badly mutilated that they were virtually fused together. Paddy Hayden from Ballyfermot told the *Irish Times* of the man he saw lying on Parnell Street screaming in pain, his spinal cord smashed. For emergency workers, any vehicle in the area immediately became a *de facto* ambulance such was the carnage, part of a fleet which carried the dead, the dying and the injured. Scenes of chaos met ambulances and other emergency vehicles as they arrived at hospitals. The closest to both Parnell Street and Talbot Street, Jervis Street hospital, was virtually cordoned off by gardaí as they freed up the surrounding streets to give the ambulances multiple access routes.

Frank Kilfeather of *The Times* said that the scene in that hospital was one of pure anguish and outrage. 'The dead were removed to a special section of the hospital. The injured were quickly examined by doctors and given labels to work out a system of priority.' As the doctors triaged the scores of arriving victims, relatives and friends congregated in the corridors and wards, awaiting news. Dr Anthony Walsh told relatives, 'We are sorting out that information as fast as we can. As soon as we have positive information, we will give it to you.' *The Times* journalist also says that, 'the doctors and nurses carried out their work with a discipline, kindness and understanding which merits recording.' One of the witnesses on Talbot Street that day was an anaesthetist from Belfast's Mater Hospital. He was only thirty yards away when the bomb exploded. 'The scene was horrifying,' he told the *Irish Independent*. 'I ran back to see a woman on the pavement decapitated; another woman lay dead with a piece of car engine embedded in her back; a man was dying with an iron bar through his abdomen. There were injured people all around and many of them could have had but a fifty-fifty chance of surviving. A woman held a seven-month-old baby covered in blood but I was able to assure her the baby was uninjured.'

Little mention is made in the press of the efficiency of the emergency services; the *Irish Press* is probably the most comprehensive. On 18 May it describes the integral part that the 'Red Alert Emergency Plan', formulated in the late Sixties, had in the swift emergency response to the catastrophe. Dublin Fire Brigade's role was to initiate and coordinate the response to the attack, which involved the usage of 'well over twenty ambulances,' according to a DFB spokesperson at the time. 'As well as the bomb victims we still had to deal with the normal run of traffic accidents and other ambulance calls,' he added. The *Press* is probably the only national paper of the day which tries to convey the human face of the emergency services and the work that they did that day. The paper reports on the work of fifteen-year-old Esma Crabbe, a volunteer with St John's Ambulance, who was being given a lift home in an ambulance when it was diverted to Parnell Street. The young girl was soon trying to administer aid to a man lying motionless beside a car. 'I took his pulse and he was dead.

Then I was called to a man covered by a plank. When I lifted it up one of his legs was missing and lying nearby. One side of his head was completely ripped away and was lying on the ground . . .' From a Dublin Fire Brigade perspective, an exceptionally sad picture on page five of the same day's paper conveys the hardship that the city's firefighters experienced that day. Three DFB men, visibly upset, are carrying a dead baby wrapped in a blanket from a building. On the same page, an on-scene reporter describes a fireman struggling to hold back the tears.

The victims of the explosion on South Leinster Street are often overshadowed by the greater carnage on the north side of the Liffey. A journalist with the then *Cork Examiner* wrote: 'Windows in several blocks of buildings on opposite sides of the street were blown out by the terrific impact of the blast, which shook buildings in all the surrounding streets. Rivers of blood mingled with the sea of broken glass that spread along the roadway.'

The *Sunday World*, Ireland's only tabloid at the time, led its paper the following Sunday with a horrific image of Dublin firefighters tenderly carrying the burned and mutilated body of a child killed in the bombs. The body had been found beneath that of a man on Parnell Street. The child was part of the O'Brien family which was wiped out entirely in the attack. While the image drove home the savagery of the attack, it was met with revulsion by many members of the public. It is however, one of the few colour photographs of the attack.

The tallies of dead and injured mounted as dusk gathered on what the Taoiseach of the time, Liam Cosgrave, described as Ireland's darkest hour. Speaking on RTÉ television he said, 'It is in moments of greatest tragedy like today's that we see and appreciate the dedication and unselfish service of the gardaí, army, ambulance and fire services.'

Tommy Ellis remembers being relieved after 8 pm that night; his shift had just been due to finish at 6 pm when the bombs went off. 'There are parts of that day which are a blur but I remember I was very distressed and it only began to sink in when I got home. Of course, my wife was frantic with worry, she knew that I would have been involved in the response to the explosions and had no idea if I was OK or not.' The following day offered no respite from the trauma. 'The next day was even worse for me. We had to return to the morgue in Jervis hospital and identify what we had brought in. It's strange, especially since it was such a large-scale incident, but I remembered every dead body or body part which I had recovered from the scene. I knew them the second I saw them. The whole thing was incredibly upsetting and I have to say it was one of the toughest working days I can remember.'

Meanwhile, just ninety-three minutes after the devastation had been triggered in the capital, further horror was visited upon the town of Monaghan as a massive bomb in the centre of the county capital exploded at 6.59 pm outside Greacen's bar on Church Square. Five died instantly and over twenty were injured. Fire and ambulance crews worked feverishly through the mayhem alongside the gardaí as a pall of black smoke gathered overhead, near to the massive crater left by the blast. Five of the victims were either working or drinking in Greacen's Pub at the time of the attack. A sixth died waiting outside his car on the opposite side of the road. The bomb completely destroyed the town centre, glass shards and shattered debris rained down over a 200-yard radius after the blast.

One of the local newspapers, the *Northern Standard*, praised the work of the emergency services in the aftermath of the atrocity. Local firefighters, ambulance personnel and civil defence members joined the gardaí on the front line to care for the injured and to retrieve the dead. Meanwhile, firefighters were also faced with a number of relatively major and minor fires in the immediate vicinity with cars ablaze and the nearby Central Café burning under a cloud of heavy smoke beside the badly damaged Hibernian Bank. An excerpt from the *Northern Standard* of the following Friday, 24 May: 'So expeditious was the rescue operation that within ten minutes of the blast all the injured and the dead had been removed to Monaghan County Hospital. The ambulances had the assistance of passing motorists who readily offered their assistance in transporting the injured. Stunned and taut-faced as they were, the rescuers, whether members of the emergency services or volunteers worked with an urgency and efficiency that excited special praise from all and won official praise from members of the Monaghan Urban Council at a special meeting on Saturday afternoon.'

The bombs in Monaghan and Dublin remain the greatest post-War atrocities in the Irish state. In the aftermath newspaper editorials heralded a dread chapter of death and destruction on the southern side of the border, a premonition which thankfully remained unfulfilled. However, for families and friends of those killed in the atrocity the lengthy process of waiting for justice was just beginning, waiting for answers as to what happened that awful Friday. In March 2004, the Oireachtas Joint Committee on Justice, Equality, Defence and Human Rights published its own report into the atrocity and its aftermath. Amidst the 119 pages of the report, written as it is for submission to the houses of the Oireachtas, some of the testimonies of those who survived the bombings but have had their lives shattered provide a strong case for a fresh investigation.

The Sub-Committee stated in its report that the 'true cost of these atrocities is incalculable.' Ms Michelle O'Brien, speaking to the committee, lost her mother in the attack. She told the committee about the actions of her father who arrived home from work to find his wife missing: 'He started to search the hospitals and in the early hours of Saturday morning he found her remains in the morgue. He knew it was our mother because she had worn a green coat and by her wedding ring which I am proud to wear today.' Kevin O'Loughlin, whose mother died on South Leinster Street, told the committee of the unbearable agony as he waited for his mother to come home safely from a city in chaos. They knew that her route home took her right through the scene of the explosion, and he recalls his father finding her body in the morgue that night. Due to the nature of her injuries, the rest of the family were not permitted to see the body. 'I did not see my mother's body when she was killed and I have no memory of what she looked like. She was wiped off the face of the earth in the memories of me and my brother. One day she was there and the next she was gone.'

For those who dealt with the atrocity head-on, the memories were engrained forever. In the aftermath, many suffered terrible mental scarring. The *Evening Press*, 31 May, carried a special feature entitled 'What Price the Heroes', written by Clare Boylan. It was one of the few pieces at the time to deal with the effects of the bombings on gardaí, nurses and firefighters. An excerpt: 'Last Sunday week a man sat down to his Sunday dinner in the house of his friends and broke down and wept. The man was a Dublin fireman, one of the heroic men in uniform who sorted through the litter of life and helped to save scores of lives. When his friends began to talk about the bombings, he started to cry.'

In typical firefighter style, most of the quotes are unattributed, but the depth of the bombing's effect on the psyche of the emergency services was clear. 'A lot of people think we're immune to emotion and tragedy. It affects us just like anyone else, except we have to get the job done before we can let it show. Our feelings get the brunt of it. None of us has been able to sleep since that night because the picture is always in the front of your mind. Sometimes I wonder why I ever joined the brigade because at times it can be a bloody awful job.'

Dermot Ryan worked with Tommy Ellis and Denis McGrath on that terrible day and he is pictured alongside them in this article. He described what it was like to try to comfort those who had lost so much in the devastation. 'They sat and cried, you try to comfort people, you'd say things like, "you're all right now, we'll look after you." But so many of them just weren't going to be all right.'

And so it was for the firefighters. Even today, Tommy Ellis says he lies awake and thinks of some of the things he's seen during his time with the brigade, and that day in 1974 is never far away. 'It's the worst memory I have and it'll never go away. I can still see most of it when I close my eyes or when I lie awake some nights. After the incident I was distressed but at the time I didn't realise exactly how traumatised I was. It was the same for many of us and I know I'm not the only firefighter who still has memories and nightmares of that day.'

The victims of the Dublin and Monaghan bombings are:

Monaghan: Patrick Askin (44), Thomas Campbell (52), Thomas Croarkin (36), Archie Harper (73), Jack Travers (28), Peggy White (45), George Williamson (72)

Dublin – Talbot Street: Josie Bradley (21), Anne Byrne (35), Simone Chetrit (30), Concepta Dempsey (65), Collette Doherty (21) & her unborn baby, Breda Grace (35), May McKenna (55), Anne Marren (20), Dorothy Morris (57), Marie Phelan (20), Siobhan Roice (19), Maureen Shields (46), John Walshe (27)

Parnell Street: Marie Butler (21), John Dargle (80), Patrick Fay (47), Elizabeth Fitzgerald (59), Antonio Magliocco (37), O'Brien Family; John (24), Anna (22), Jacqueline (17 months), Anne Marie (five months), entire family wiped out in attacks. Edward John O'Neill (39) and his stillborn daughter, Martha, born in August 1974. Breda Turner (21)

South Leinster Street: Christina O'Loughlin (51), Anna Massey (21)

Central Hotel Fire

'Mesmerised faces, some of them tear stained, gazed at the
blazing building of the central hotel as the relentless and
uncontrollable flames roared through it, lighting up the street in
a fiery glow and sending sparks billowing high in the sky from
the inferno ...'
Vera Cave, *Donegal Democrat*, August 1980

Ten lives were taken in the fire at the Central Hotel on 8 August 1980,
at the height of a long hot summer in Bundoran, one of the North
West's most enduring holiday destinations. The fire which ripped this
building from the heart of the town, decimating families and scarring
the lives of many survivors, remains etched in the memories of the
firefighters who faced it. It remains the single most destructive
inferno ever faced by Donegal's fire service. Retired s/o PJ Clancy
remembers the intense heat which caused plastic signs along the street
to melt, dripping vivid colours onto the pavement below.

The day after the blaze, nine bodies were retrieved from the debris
of the once-proud, graceful 19th-century premises. The flames left
very little for the families of the bereaved. For the family of eighteen-
month-old Nicola Lamont they left nothing at all.

August is traditionally the boom month for tourism in towns such
as Bundoran and over 25 years ago a robust evening's entertainment
was coming to an end in the bar and function room of the Central
Hotel. There were sixty guests booked into the hotel's 45 bedrooms,
mostly a mixture of families and couples. James Warnock was a
barman in the hotel at the time and was just closing up for the
evening when he noticed smoke in the reception area of the hotel. 'I
saw flames at the door of the bar,' he told the local papers the
following day. 'I let out a roar and yelled to the people to get out
quickly. When we got into the lounge the flames were going through
it. It was a massive blaze. Outside, it was total chaos. I saw people
being thrown out the back windows. Anyone who didn't get out when
the screams went up didn't have a chance,' he said. James Warnock
and the assistant manager of the hotel were the last to get out to safety.
'If I'd stayed two minutes longer I wouldn't be here today.'

Meanwhile, members of Bundoran Fire Service were arriving on
scene. Retired driver Johnny Mulreany remembers the callout. 'It was

before we had pagers of course, so the first we knew of the fire was when the siren sounded from the station. I live at the far side of the town and had to drive by the fire to get to the station. I remember seeing a burning car on the footpath from a distance, and thought it wasn't so bad, then I got closer and could feel the heat from the flames inside the hotel.' At the station, the crew prepared their appliance, an ex-army Bedford 'Green Goddess' fire tender.

PJ Clancy, who lives just beside the station, remembers seeing the inferno over the rooftops when he was on his way to the station. 'I knew then that this was serious, very serious indeed.'

Meanwhile, down at the Central Hotel, there was pandemonium on the streets as locals rushed to the blaze with blankets and in some cases ladders, pioneering a makeshift rescue effort as the fire crews were arriving on scene. Blankets were held out as makeshift safety nets — as the inferno ravaged the hotel interior, leaving windows as the only escape route. It was the only choice for the McCombe family from Strabane. As heat rose from the flames below, the floorboards of their second-storey room began to creak with the heat. 'We could either jump or be burned, the room was already filled with choking smoke' said Kathleen McCombe. Her mother, Marcia Edgington, was the first to leap from the window, falling the twenty feet in dread fear. However, vigilant locals on the ground intervened, placing a blanket beneath the falling woman and minimising her injuries. Kathleen then had to simply drop her terrified child Robert, screaming in fear, from the window before jumping herself. While she landed on the blanket, her son was not so lucky and missed the makeshift safety net completely, sustaining severe injuries in the process. All three were rushed to Sligo general hospital for treatment.

The ordeal of the McCombe family was one of the night's more fortuitous dramas. The Brennan family from Naas had checked into the Central Hotel not long before the fire broke out. James Brennan and his wife Anne, along with their two children Deirdre (six) and John (three) were in Bundoran for a short weekend break. Staying in the Central Hotel on the recommendation of friends in the area the family were in their rooms when the blaze engulfed the hotel. All four perished. 'I remember the next day; we discovered the whole Brennan family in the reception area of the hotel. The floor of their room had burned through and come crashing down,' recalls PJ Clancy.

The Kinsella family, from Artane in Dublin, had travelled to

Bundoran to celebrate: Jimmy Kinsella had just got a new job with the Post Office and along with his wife Joan they had taken their four children Jennifer (16), Geraldine (16), Adrienne (13) and James (nine) to the Donegal seaside resort on a spur-of-the-moment break. It was the first night of their impromptu holiday when the tragedy at the Central Hotel claimed the lives of Adrienne and James despite desperate efforts to rescue them. One of the older children, Jennifer, only survived by jumping from the inferno. PJ Clancy remembers Mr Kinsella pointing desperately to the window of the room where his children were, begging the firefighters to intervene. 'I remember putting the ladder up to the window and going up. But inside, you could see nothing. Absolutely nothing, just flame and smoke. Every window and door was engulfed with it, the rooms were destroyed. He tried to get us to go in several times and even tried to scale the ladder himself. We had to stop him, of course, anyone who entered that fire would never come out. I'll never forget the condition of the poor man. I know that if it were my children inside that building, I'd do exactly the same.' The two children were later buried after a funeral mass attended by a crowd of around 3,000 in Beaumont, Dublin.

As the flames destroyed the hotel the crowd which gathered to witness its destruction included Eamonn Dowdican, now a firefighter with Bundoran Fire Brigade. Then a young boy, Eamonn was unaware at the time that the only local victim of the tragedy that night was his Aunt, Sadie Dowdican, who had been working as a housekeeper at the hotel. 'I remember standing in the car park behind the hotel, watching the whole structure fall to pieces. From what we had heard from the crowd we thought that the victims were tourists; I didn't know until the next day that Sadie was amongst the dead.' Ms Dowdican perished in the top storey of the hotel, directly above the room of Pauline and Nicola Lamont. A single mother, Pauline escaped from the flames but was unable, despite frantic efforts, to save the life of her dear child, who was only eighteen months old. Some days later, she said a poignant farewell to her baby when she laid a single red carnation in the blackened shell which had once been their hotel room.

PJ Clancy says that the response to the fire was swift and that nothing more could have been done because the fire moved with a malicious speed which confounded any serious rescue attempts. 'We were there quickly, within eight minutes of the alarm, but the flames were so fierce and so fast that within an hour all that was left of the

Central Hotel was a smouldering ruin.' The Bundoran brigade was assisted by other crews from Ballyshannon, Donegal Town, Manorhamilton and Sligo. *Donegal Democrat* journalist Vera Cave recalled the scene in the following day's issue of the newspaper:

'Windows on the other side of the street cracked with the intense heat; firemen valiantly set up their equipment and hosed optimistically at the stonework, steam hissing as the water hit the hot bricks. Ambulance crews tended huddled survivors, gardaí moved about in the disorder urging the tense, babbling crowds of onlookers back; hoses snaked along the street in puddles of water and the crowds grew ever bigger as more and more people arrived, attracted by the smoke and orange flames visible for miles. The flickering light and the helmeted firemen were reflected in the water-drenched streets as they fought to prevent the fire spreading to the adjoining bank and locals joined with strangers in efforts to get ladders up to the building and save any lives that might be left. The people, eyes fixed on the hotel from whose paneless windows came gusts of fire from the brilliantly lit interior, talked among themselves, speculating on the cause, the effects, the survivors and the victims. In a close-knit mass, they swayed back in panic when girders cracked or fell or when the ominous sound of falling bricks reached their ears. Sometimes a red-eyed, sobbing girl or parent pushed through the people searching for relatives or friends, and eventually, as the danger of the building's collapse became imminent, the crowds were roped off on each side of the furnace, only to crowd as near the barrier as possible once again.'

Johnny Mulreany recalls the intense heat of the blaze. 'Entry into the building was never an option for any firefighter. The streetscape around the hotel could give you an indication why. Large glass windows in every building near the hotel cracked and broke with the searing temperatures and, as PJ has described already, plastic signs within a significant radius melted away.' As the first shards of dawn loomed over Bundoran on Friday morning, weary firefighters, who had fought the fire all night and continued to damp down the blaze, now faced the most gruelling task of all. It was about 8 am when firefighters and ambulance personnel got the order to commence the recovery operation and enter the hotel. 'As firefighters, when we entered that hotel, we needed to be in the proper mindset. There was a job to be completed and we set about getting it done,' recalls PJ Clancy. 'A scene like the interior of that hotel is something which I have never seen before or since, but like many other tragic events I and other firefighters have attended, you need to detach yourself to a certain extent and complete the task. You can't think of anything else and, to be honest, in the majority of cases, you don't have time to think of anything else. I'll never forget what I saw in that hotel that morning though, I still see it and smell it to this day.' Within a matter of hours of the last of the flames being quenched, the bodies of six of the victims of the inferno had been retrieved. Later that morning, a seventh was discovered. The recovery operation continued into the weekend with firefighters, ambulance personnel and gardaí carefully sifting through the debris. The eighth body was retrieved on Saturday and the ninth on Sunday. No trace was even found of the tenth victim, Nicola Lamont. 'Every grain of rubble from that fire was sifted through. We never found anything of the child,' says PJ Clancy.

After the terror and turmoil of the blaze came the investigations and recriminations. There were those who insisted that the hotel could have been saved if the fire brigade had had the correct water pressure. Dáil Deputy Jim White said, 'I have at least ten sources all very critical of the water supply. I am not saying that the hotel could have been saved, but if there had been a storm blowing that night the whole street could have gone.' His claims were refuted by Donegal County Council and the Department of the Environment. Council Chairman Leo McGovern was also angered by headlines in the *Irish Independent* on the Saturday after the fire which claimed that low water pressure had cost lives. While the firefighters agree that the blaze presented a serious danger to neighbouring buildings, they don't agree that more water or increased water pressure would have made any difference although they concur that the water pressure system at that time wasn't what it could or should have been. Such ruminations, the firefighters maintain, are ultimately futile, though, once the intensity of that blaze is considered. 'You could have hit that fire with the Atlantic Ocean and it wouldn't have saved that building or those people,' says PJ Clancy. 'Over the years, I have never had anyone in this town say to me that we could have done something that would have changed the course of events that night.' Other theories which abounded in the days after the tragedy included arson and the use of incendiary devices by Loyalist paramilitary groups from across the border. In a statement, gardaí said that there was no evidence of incendiarism and that the initial stages of the investigation pointed to an electrical fault having triggered the blaze. That is still the accepted cause to this day, with faulty wiring in a television set the suspected culprit.

On Friday night, after 23 hours on the fireground, battling a fire and recovering victims' bodies, PJ Clancy finally made his way home. Physically and mentally, he had to deal with the aftermath. 'I was obviously shattered tired and my mind was racing backwards and forwards with everything I had seen, but after a few hours' rest I had to go back to the fireground so I really didn't have time to think about it. If an alarm went off that following night or the day after we all would have got in the appliance and responded to it. There was no comprehension or investment in traumatic support services. We just got on with it.' Johnny Mulreany said that if a member of the crew had taken the events of that night particularly badly, he was expected to deal with it himself. That's just the way it was. 'It's the way it still is in some ways,' he adds. 'Firefighters tend to keep to themselves.' The fire at the Central Hotel devastated Bundoran for years, the shell of the building left as it was for over a decade, an open wound to the events of that night. Today, another hotel, also called the Central Hotel, stands on the same location. As of yet there is no plaque or other symbol to commemorate the ten victims who died there over twenty-five years ago.

Cherryville and Buttevant

It was 5.28 pm on a Sunday in August 1983 when the train from Tralee to Dublin got under way. As always it was packed full for the trip back to the capital at the end of a weekend. It was behind schedule from the start and problems were noticed before it even reached Mallow, where it was swamped with even more passengers. The engine had overheated and had to be replaced. In the interim a similarly busy train had left Galway, also bound for Dublin, at 5.18 pm. Just after half-seven, this train was hurtling towards Athlone after leaving Ballinasloe bang on schedule. The Tralee train, meanwhile, finally departed Mallow after a half-hour delay. It had a new locomotive, but it had not been supplied with enough fuel to get it to Dublin.

After 9.20, the Galway train had passed through Portarlington and was less then 15 minutes from Kildare Town when it was stopped by Central Control in Heuston. The Tralee train was running over half an hour late and Central Control at Connolly Station instructed that it should be allowed onto the main Dublin line first — it was normal practice to minimise delays on an already-late service. The train from Kerry went through onto the main line but briefly before 9.30 it ground to a halt outside Kildare Town, at a place called Cherryville Junction. Tired and agitated passengers were told that the locomotive had run out of fuel. There were almost forty people standing towards the rear of the train, approximately twenty in the dining car. The driver began to walk along the track to a telephone to update Heuston station on the situation. It was late on a summer's evening now, and the track was thick with fog. On both sides of the stranded train lie nothing but marsh-laden countryside.

One of the guards from the Tralee train set out to lay warning detonators for any following trains. However, the Galway train had now entered the mainline and was heading towards Dublin, unaware of its stationary counterpart ahead. In 1983 there were no direct radio links to the trains, drivers instead being instructed by track signals. The Galway train was moving slowly through the thick fog, to the annoyance of many of its passengers who by now were sitting in virtual darkness in some of the unlit carriages — it was now 9.32 pm.

The guard setting the warning detonators for the Tralee train had set two when he heard a rush of noise and saw another train looming out of the mist, not at great speed but too fast to stop suddenly. The Galway locomotive smashed into the rear of the prone train ahead, obliterating the guards van and lifting the standard carriage off its chassis. It then thumped into the rear of the dining car which ended up strewn on an embankment to the side of the rails, pointing twenty feet into the air. Lights failed on both trains and everything fell into darkness. Panic reigned. It was 9.40 pm.

With the Galway train travelling at approximately forty miles per hour, this train accident claimed seven lives and inflicted almost thirty serious injuries. If the Galway train had been travelling at greater speeds the scene which awaited rescue workers could have been far worse then that at Buttevant, three years previously.

As the Galway train bore down upon the stationary carriages

ahead, Dermot Nash (35) was talking to a fellow passenger when through the window he clearly saw the other train ahead in the gathering dusk. Just before the crash, he looked at his fellow passenger and said, 'Jesus, do they not have radios on these trains.' He was thrown against the wall by the impact, before struggling to his feet and crawling from the carriage. To his right he saw three bodies, lying together in a group. Ahead, at the epicentre of the tragedy, it was pure carnage.

John Byrne, a local priest, was one of the first upon the scene after the rescue workers. While emergency crews went to work, Byrne could appreciate the devastation of the scene for a moment. 'It was unbelievable,' he told reporters. 'One carriage was up in the air, it was just a tangle of metal.' Then Byrne was moving amongst the injured and dying, within minutes he had administered the last rites to four people. Such was the force of the crash, the two rear carriages of the Tralee train had been driven up on top of the Galway locomotive before pitching to the side. When the trains came to rest, passengers frantically tried to escape from the wreckage, finding themselves in total darkness, surrounded by fields and marshland.

Kildare's CFO, Michael Fitzsimons, had only been in the job a few short months when he was faced with this tragedy. 'The main problem was that the accident happened in the middle of a bog in the dead of night. Most passengers had no idea where they were and in a confused state made their way towards the main road. In a train crash it's often quite hard to determine exactly how many people were on board at the time. There was a lot of confusion just trying to establish the number of injured as many of them had been taken away by private cars.' For today's firefighters, imbued with superior equipment and a certain degree of resilience given the frequency of nasty RTAS, the train crash would have been less challenging. 'Dealing with disasters of this kind is part of the job but that doesn't mean you're not affected by it,' said Fitzsimons many years later. 'It's an incident I still think about, you have to be mentally prepared for something like that, knowing that a similar incident could happen tomorrow.' Indeed, similar incidents had happened that very day. There were two other derailings that day, one in Kildare, at Moyvalley where a train equipped to carry chemicals had become derailed. If the train had been carrying its normal cargo of explosive Acrylonitrile, the county's emergency management plan would have had to have been put into action twice.

Down at Cherryville Junction in the early of hours of Monday morning, the extent of the crash was being explored at the gruesome epicentre where the trains had collided. Pointing like a stricken finger into the night sky, the dining car had spilled its contents around the ground. A member of the Civil Defence, who had been working with firefighters and ambulance personnel to clear the injured and the dead from these carriages, pointed at the carriage and told Denis McClean of the *Irish Press*: 'All you can see is a hand and we don't even know if it's that of a man or a woman.' Around the carnage, firefighters with torches scoured the marshes for further remains or those who may have been thrown clear by the impact. The job of clearing the dead from the last two carriages of the Tralee train had been particularly harrowing. It was around 3 am when the last of the dead or seriously injured had been extricated. The bumpers of both trains were smeared with blood and Newbridge s/o Patrick Mockley remembers the body of the final victim, a member of the train's staff. For firefighters, the body wasn't hard to find; 'We saw his blood dripping from the wreckage, he had been very badly trapped and it took a long time to get him out. The whole scene was horrendous. It was surreal to see a train in that condition and also quite traumatic.'

The following morning revealed the true horror of the crash. Maurice Walsh, writing on scene for the *Irish Times*, described the scene: 'As light dawned the horrifying dimensions of the crash became horrifyingly clear. The dining car of the Tralee train, its wooden walls in pieces around the track and its shell in the bushes, stood ruptured and broken. Jagged pieces of steel white where connections had been snapped off. All around the track lay artefacts of an earlier normality. The track was strewn with plastic spoons and sugar sachets from the dining car. Here was a blue sleeping bag, there a navy jumper. There was a soiled but untorn copy of *Going Home* and not far away a dirty £10 note. Further down the track were reminders of how that normality had been broken; an odd blanket and bloodstained tissues.'

Just over three years before the terrible accident at Cherryville Junction, eighteen people had been killed in Buttevant in Cork when a train derailed while travelling at seventy miles per hour on 1 August 1980. A further sixty-two were injured in what remains Ireland's worst rail tragedy. The 10.30 am train from Dublin to Cork, carrying over

230 passengers, derailed as it approached the station at Buttevant. The tragedy happening in early noon, its horrific scale was apparent immediately. In the press, it was described as akin to a battlefield. When the train had derailed, it had slammed into an embankment alongside the tracks which led to the carriages behind jack-knifing across the tracks. The first three carriages bore the brunt of the impact, shattering them into sections, causing terrible laceration and crush injuries to those inside.

Nurses clad in white were some of the first on the scene from nearby Mallow hospital, as ambulances and fire engines from Mallow, Charleville and Cork City began to gather. It was an extrication and recovery exercise for firefighters. One victim, in severe pain from injuries, was attended to by a nurse who was lowered down beneath the carriage by two firefighters to administer a morphine injection. Although many of the injured were screaming for aid from inside the carriages, firefighters and rescue personnel had great difficulty in reaching them because of the distorted wreckage. Heavy lifting gear arrived later and the afflicted carriages were lifted and placed in a neighbouring field, so they could be thoroughly examined by rescuers. It was a massively traumatic day for the country and indeed for Cork, coming the year after the nightmare of the Whiddy Island explosion. The incidents at Buttevant and Cherryville are the third and fourth worst rail disasters in Irish history. In 1889, eighty people were killed in a crash near Newry, County Armagh while in 1945, twenty-two were killed in an incident at Ballymacarrett, near Belfast.

Raglan House
In 1987, Dublin City was rocked by two serious gas explosions in the space of less then two months. The explosion at Raglan House, a block of apartments in Ballsbridge, left two people dead and rendered scores homeless in the first hours of 1987. Prior to the explosion, ten gas leaks had been reported by the residents to Dublin Gas Company. In the days leading up to the explosion, two more terrible acts of negligence contributed to the tragedy. On Christmas Eve, Mr and Mrs Michael and Noëlle Murphy again noticed a strong gas smell and reported it to Dublin Gas. The company called to the flat and despite carrying out some work which reduced the gas levels outside the flat, they made no measurement of the gas levels inside the Murphy house itself or found the leak which was causing the build up. The problem

was left over the Christmas period, but those remaining in Raglan House continued to notice the strong presence of a gas smell. On the day prior to the explosion, 31 December 1986, the area was again examined. Again, measurements were taken outside by the Dublin Gas Company, reading similar levels to those of Christmas Eve. It was decided that the problem was not serious and could wait until the following day due to the volume of calls which the company was receiving. The crews left Raglan House.

The next morning, at around 8.30 am, a resident of Flat 2 in Raglan House awoke and noticed the now-familiar smell of gas, though this time in her hallway. She checked the pilot light in her kitchen, saw it alight and then prepared to leave the house after breakfast. The first explosion occurred at 9.30 am. It ripped upwards through the face of Raglan House, completely obliterating flats numbered 1, 5, 9, 13 and 17. The debris of these dwellings crashed to the ground, while behind them the remaining flats remained mostly intact. The residents of flat 5, Michael and Noëlle Murphy, were killed in the explosion and succeeding collapse. Both worked in RTÉ and one of their colleagues, Sean Kelleher, also lived in the block: 'I thought it was a local explosion, something that happened in a neighbouring apartment. It blew in the door of my apartment and the shock wave blew out a window in the lounge.'

In the subsequent investigation into the event, commissioned by the Minister for Energy, it was surmised that the pattern of the first explosion originated in a gas air explosion in flat one. The report into the tragedy also surmised that the trigger for the explosion was the switching off of the central heating timer at some time between 9 and 9.30 am. Of course, the explosion bred fire. One of the residents of the flats which survived the initial explosion said that she remembers the plasterboard of her bathroom wall collapsing onto her in her bedroom and noticing an orange glow. She escaped through the bedroom window to the rear of the building. Meanwhile her son was in the adjacent bedroom. He had been awakened by the plaster board from the hallway collapsing on top of him. Not seriously injured, he crawled out of the debris, dust and smoke which had swamped the flat and made his way outside to safety.

Ms Gráinne Murphy, another resident, said that she had heard a sucking noise before her bedroom window shattered: 'I thought it was a bomb in the (nearby) bank. I heard people screaming, I came downstairs. There was a horrible smell of gas.' Local baker, Denis Lumley, told reporters that he thought the roof of his shop had exploded; 'It seemed to jump. I ran around to the front. There was no sign of anybody, I didn't hear anyone shouting. I went back in and dialled 999. Within a few minutes, squad cars had arrived.'

The nearest DFB station was Donnybrook and the first 999 call was logged at 9.32 am. The first fire appliances began arriving on scene within the next eight minutes. The firefighters emerged from their vehicles to find a fire in the ground-floor area of the flats which remained, with sporadic ignition in the debris of the first explosion. The fire in the structure was quite intense, with plenty of easy-burning household elements to fuel it. Eight fire appliances fought the blaze while firefighters used thermal imaging equipment to search the rubble for victims. With the scale of the incident, the Major Emergency Plan for Dublin was rolled into action. The three main hospitals for the incident were St Vincent's, the Mater and the old Richmond Hospital — extra surgeons were called on duty and medical specialists were despatched to the scene should they be required.

Between ten and twenty minutes after the arrival of firefighters and other emergency workers a second explosion rattled through what remained of Raglan House. This second detonation didn't have the raw power of the first, described more as a muffled bang. It did, however, lead to serious structural deterioration in the remaining flats, with two more balconies joining the debris pile below. The Dublin Gas Company had shut off the main gas valve to Raglan House just after 10.15 am with the assistance of DFB fire officers. Throughout the morning, it became obvious to emergency workers that the remaining flats were likely to collapse, as they observed the widening of cracks in the unsupported upper storeys; firefighters were withdrawn in time before the eventual collapse which occurred later that morning. According to Garda Michael Toher in *The Irish Times* the following day: 'I was close to the building and there were a lot of firemen around. Suddenly there was loud noise and everyone instinctively jumped back. Fortunately, the debris did not fall outward or there would have been many people killed. It was the most amazing escape I have ever seen.'

In the afternoon heavy lifting machinery came on site to assist the firefighters with the extrication work of any possible casualties. It was then that the remains of Mr and Mrs Murphy of flat 5 were uncovered. In the aftermath of the event, one of the first on the scene was former Taoiseach Garret FitzGerald, pictured being briefed by DFB Acting Chief Fire Officer Michael Walshe. The brigade itself did not escape totally unscathed from the drama that morning with firefighter Gerard Mulligan taken to hospital after being injured during the search for victims.

The Cremer and Warner report in the wake of the explosion at Raglan House and later Dolphin House severely criticised the safety standards of Dublin Gas, castigating the company for having no overall safety policy. In the aftermath of the tragedy in Dublin 4, Bernard Somers of Dublin Gas Company admitted that customers had the right to be uneasy but there was 'no cause for alarm'. While the aftermath of the Raglan House tragedy led to no rapid change, the destruction in Dolphin House less then two months later led to Dublin Gas Company conducting a serious survey of apartment blocks in the city. The destruction at Dolphin House was minor compared to that in Ballsbridge but it still left one resident with serious burn injuries after three internal walls in the Corporation building were destroyed. The Cremer and Warner report found that the explosion was triggered by an electrical device such as an electric fire, but added that the solid construction of that building avoided further collapse such as that which had claimed two lives in the tragedy at Raglan House that New Year's Day.

Part Five
Saying Goodbye

The trauma of saying goodbye to a colleague in any job is an upsetting one and firefighters do not seek to put more stock in their loss then in that of others. Nevertheless, like other emergency services, the fire service places a group of people in a situation and an environment where, by its very nature, close bonds are forged. Virtually every station in the country has had to say goodbye to a colleague, whether taken quickly in the course of their duties or through illness or the natural course of time. The notion of the 'firefighter's' funeral, so poignant since the events of 11 September 2001, has become synonymous with heroic death in the line of duty and self sacrifice.

The firefighter's funeral transcends the barrier of how death occurred. It doesn't matter. There have been funerals for those who died in the line of duty that have been matched equally with those of a firefighter who died as a result of illness or a freak accident. In a brotherhood as close as that in the fire station, it's not just how well you did your job but how well you lived your life that is remembered by your colleagues.

The last fire which claimed the lives of DFB firefighters in the line of duty was in 1936 when firefighters Tom Nugent, Peter McArdle and Robert Malone were killed battling a blaze in the Exide batteries factory in Pearse Street. The fire occurred on 5 October 1936 and the three perished following a massive explosion shortly before 11 pm. In the aftermath of the tragedy, the city's debt to its firefighters was recognised as the three bodies lay in state. The condition of the city's water mains, which many blame for the tragedy, was also highlighted. The three firefighters were buried in Glasnevin cemetery. Today, a modest plaque marks the spot where they died in the line of duty, though there are plans for a more fitting memorial to be erected.

Eight years earlier, a County Cork firefighter, Michael O'Connell from Kilmallock was killed in a major fire in the early hours of 16 May. As the town itself had no brigade, when the fire at the building of Messrs Cahill & Co. raged out of control, the firefighters from the city of Cork were summoned to assist. When the firefighters arrived, they successfully fought the fire and stopped it spreading to neighbouring premises. With the blaze knocked down, the firefighters were in the process of mopping up the remains of the fire when, without warning, the front wall of the building collapsed, burying firefighters Michael O'Connell and Denis O'Leary under the rubble. When they were dragged free by their comrades, they were treated at a local hotel by doctors before being rushed to Croom hospital, both with severe injuries. O'Connell was critical and his family were summoned to his bedside, his wife and his three young children. While O'Leary survived and recovered, O'Connell died on the evening of 17 May, aged 31. After his funeral, O'Connell's family emigrated to America where the tradition of their dead father was continued; today his grandsons work as firefighters in Dunkirk, New York.

Willie Bermingham

While to be a firefighter is to encounter the human condition at its most vulnerable there are few, however, who dedicate so much of their lives to ending human suffering and misery as the late Willie Bermingham. A dedicated firefighter, in the course of his duties Willie came across a hidden underbelly of human suffering in Dublin which compelled him to establish a charity to assist the elderly, weak and poor of the city. He called that charity ALONE and it is still prominently active.

Willie first wrote about their plight in the Christmas 1977 edition of *Brigade Call*. In the earlier part of that year, in the course of his work, he had seen at first hand what was happening in desolate bedsits and tenements around the city:

'He was an old man in his seventies. Like many others, both men and women, in our city he was cast away on the scrap heap of old age to face loneliness, cold, hunger and misery. . . . On a cold wet day last February he went out, no-one knows for what or to where, but he did go out of his little home in the south centre city of Dublin. He returned and his clothes were very wet, so wet that his frail body was also wet and very cold . . . He lay on his bed and he must have gained enough strength to pull the clothes over him to try and keep warm in his wet clothes. Then some time later he felt he wanted to get up out of bed but he was only able to grip the bedclothes with his left hand ... How long it took for him to die I don't know. But die he did, in the most horrible way possible. I found him still clutching the blankets in his left hand and his right leg drawn up. The Fire Brigade had been called by someone who missed seeing the old man out and about. When we arrived we were told by some people who lived nearby that he had not been seen for six or seven days. My superior officer then decided to make entry to this little home. I forced the window and climbed in and, as I expected, there was another of a long list of what I term "my people" dead.'

Bermingham went on to describe the Dickensian poverty in which this man lived in 1977, driving home the harsh truth: 'Society had let this man die. All of us had condemned him to die alone in misery.' In addition to being a well-respected firefighter, Willie became the President and driving force behind this charity which sought to give a voice, dignity and hope to the elderly and poor. The memories of Bermingham, put into words by retired firefighter and renowned DFB historian, Tom Geraghty, portray a man of vast passions and an unquenchable desire to see the right thing done for ordinary people.

Following Willie Bermingham's death in 1990, after a long battle with cancer, Geraghty paid tribute to him in *Brigade Call* magazine. In the introduction to his tribute, the author used a phrase coined by Willie's close friend, Liam O'Cuanaigh: 'Men do not make history, but they do make their own history.' Such was Willie Bermingham's impact on the city of Dublin, that his funeral was held amidst the grandeur of St Patrick's Cathedral before his burial in Lucan. Through ALONE, he had struck a resonant chord with the poorer classes of the city, and in indeed the capital's general population, who thronged to pay their respects. For firefighters who worked with Willie, such as Tom Geraghty, they realised that they were not saying goodbye just to a colleague or a friend, but to someone who had become an institution in himself. The irony is, as was said in Geraghty's tribute, Bermingham was a man who 'abhorred opportunism, shunning it in the most direct way possible and showed this by returning money to those who gave from private or public purse in a blaze of publicity. He didn't seek charity for his people because he wouldn't allow the strong or the wealthy to salve their consciences with charitable bribes. Society had allowed his people to live in the shadow of shame and total neglect, persecuted by petty criminals and slum landlords while they suffered old age or were crippled with infirmity, so the leaders of that society were not going to be allowed get away with mere handouts.' Liam O'Cuanaigh said that Willie 'didn't give handouts; he tried to give dignity. He didn't give charity; he tried to give pride. He didn't give pity; he gave hope.'

He held this uncompromising attitude in every aspect of his life. In the fire station — he served most of his career in HQ — he was determined to remain one of the crew, fiercely loyal to his comrades. Whether it be on a fireground or at the trade union negotiating table, he was unashamedly direct. Others recognised the place which ALONE had won within the city and the clamour to reward his charitable endeavour grew. Trinity College awarded him an honorary doctorate and from the USA came the title of International Fireman of the Year. Whatever the award, it didn't change his straightforward demeanour. 'He was first and foremost a fireman, a great tribute to our profession,' said Tom Geraghty.

Delta Six-Four: Remembering Stephen 'Timmie' Horgan

'There is a strange feeling in locker room number 1 in Kilbarrack,' says Tommy Ellis. 'For some reason, many of the lockers in that room were vacated prematurely by men who died while working in the station. It's strange but we've always thought that there was a soul in the station, something watching the lads working there. If that is true, I would love it to be Timmie Horgan.'

When Stephen 'Timmie' Horgan was tragically killed responding to an emergency ambulance call on 26 August 1994, he left behind a massive void in the lives of those with whom he served. Timmie, as he was known, could often be heard before he was seen such was his larger-than-life-presence on A Watch in Kilbarrack station. Damian Guilfoyle, one of the few remaining men on A Watch in Number 5, has strong memories of his friend.

'Timmie was a big guy with a strong, boisterous manner; his voice could be heard from one end of the station to the other and whether we were on a callout or back in the station, he was at the heart of whatever was going on.' Damian joined the brigade almost ten years after Timmie, whose status on the Watch was something which he immediately warmed to and the pair became firm friends. 'In fairness, there aren't many people on this watch who wouldn't say they were friends with Timmie Horgan, he was just a tremendously likeable sort of guy. He was built like a tank and incredibly strong but also had a very generous and warm nature.'

Outside the Brigade, Timmie was devoted to his family: his wife Noeleen, sons Keith and Darragh and his daughter Ciara. He also had another passion, GAA, and was a fanatical supporter and club member with O'Toole's GAA club in the area. His brother, Bertie, was sixteen years Timmie's senior and remembers the dynamism of his younger brother. 'I was very close to Timmie but he was very different from me, he had different qualities. One of these was how he engaged and related with people. He had the knack of getting the best out of people, working or not. He wasn't ambitious in the selfish sense, he was happy as long as those around him were.' Bertie served in Dublin Fire Brigade from 1959 to 1969, leaving just before his brother joined, continuing his family's tradition of public service.

In the brigade, Timmie liked to get things done. 'Out on a job, Timmie was a force of nature,' said Damian. 'He was always in the heart of things. At a fire he didn't mess around, he just wanted to get in there and tackle it head on — that was his way.' Tommy Ellis, now retired, was senior man on the watch when Timmie joined DFB in 1972. 'I would class Timmie as a brilliant firefighter, someone who knew what he was doing inside the station and when we were on a callout. I'm not trying to make him into a saint or anything, he was just one of those guys who knew how to get things done and get on with people at the same time.'

The morning of 24 August 1994 dawned bright and hot, a perfect late-summer's day. Damian Guilfoyle and the rest of A Watch mustered for work in the docking bay of the station, ready for whatever the city might throw at them. 'I recall it was one of those days when we were all in good form, lovely sunshine. I have such terrible memories of 24 August that it's only now when I think about it that I remember what a beautiful day it was.'

On parade, Timmie was assigned to the ambulance along with Gary Burke. 'Personally, like many of us I think he preferred the fire appliance, because you're working with more people and let's face it, everyone knows you never get a minute on the ambulance,' says Damian.

They didn't get a minute that day either. Shortly after nine that morning an emergency call came through from Donabate for a seriously ill, disabled child who needed to get to hospital. The ambulance was dispatched immediately from Kilbarrack station.

'On average, responding to a medical emergency callout, you would normally be hitting a decent but not dangerous speed,' says Damian. En route to their patient, Delta Six-Four was involved in an incident close to Swords, where it is believed they struck something loose on the road. The results were tragic.

'The ambulance was heading northwards at the time. When we arrived on scene, that ambulance was lying in the ditch, facing southbound. That gives you some indication of the scale of the crash,' says Damian.

When the Mercedes crashed it flipped forward, turning over on its side and spinning into the ditch on the opposite side of the road. The crash killed Timmie Horgan immediately. Gary Burke, who was travelling on the passenger side of the ambulance, had been dragged along the surface of the road and was now pinned beneath his dead comrade suffering serious injuries to his arm and shoulder.

'6-4 is down, 6-4 is down.' When the call-sign for Kilbarrack ambulance was heard over DFB radio, Damian Guilfoyle remembers fearing the worst. 'When we heard it was our ambulance, I remember just wanting to get out there and make sure that Timmie and Gary were OK; I wasn't thinking the worst at the time.'

While Damian Guilfoyle was driving Kilbarrack's fire appliance to the scene of his friend's death, Tommy Ellis was returning from the local shops, as he was on 'mess' duty that day. 'I remember seeing Timmie leave the station on the ambulance, I went off to get the food for the day and when I returned there was nobody in the station. Of course I presumed that the appliance was off on a routine callout at the time.'

When Kilbarrack's fire appliance arrived at the crash site, Damian remembers looking into the cab of the ambulance. 'When I saw Timmie, I knew already. When you've been in the job for a while, you get to know when someone has passed on. The problem was we had to extricate Gary, who was seriously hurt but totally unaware that Timmie had died.'

Back at the station, the news was beginning to get around that something terrible had happened on the Belfast road. For Tommy Ellis, who was still totally unaware of the day's tragedy, the day was proceeding as normal before the phone rang. 'I remember that, the phone ringing and someone asking me who had died. I then went into total shock and remember just thinking about Gary and Timmie in the ambulance and just hoping that someone had got their facts wrong.'

At the crash site, the routine business for firefighters of extricating victims from a road crash was in progress except this time with a terrible twist. 'When you're using the heavy-cutting gear you have to remain focussed on what you're doing,' says Damian. 'That day, we were just concentrating on getting Gary out. We could see that he was in terrible pain. We also knew about the dark days that lay ahead when we would have to face up to the fact that Timmie was gone. Once we had got Gary out, we also had to tell him what had happened. After that I went into some sort of daze, shock maybe, I don't remember anything about the drive back to the station.'

For Tommy Ellis, dread realisation came when one of the DFB district officer's phoned to ask him if he knew where Noeleen, Timmie's wife, worked. 'I remember seeing the guys return from the callout. I knew what had happened and they knew that I knew. Everybody was just silent, distraught. Then I just couldn't take it any more and burst into tears in the locker room.' For Tommy, Timmie's death was the second death of a close friend in DFB in consecutive years, Denis McGrath having died from a medical condition the year previously.

'We were all upset, but for Tommy it was especially difficult,' says Damian. He then had to go to the Horgan household and break the news to Timmy's mother-in-law and to his 14-year-old daughter. A senior officer went to Mrs Horgan's place of work to inform her of the tragedy. 'It was like a bomb going off in that family,' recalls Tommy.

The days after Timmy's death saw the heartbreak and the camaraderie of the fire service. For the guys on the watch, support from their colleagues within DFB was ever-present and something which they always remember. 'It was great. You hear that a lot when people talk about firefighting but that's because it's true,' says Damian Guilfoyle. The day after the tragedy, there was a lethargy borne out of sadness and pure shock in the station. As senior man on the watch, Tommy was expected unofficially to lead by example. 'I was obviously as gutted as anyone, losing a comrade and a personal friend but we were still firefighters with a job to do and the listless atmosphere in the station that day was something which would have driven Timmie mad, he was so full of life. I remember saying to the lads that life and work had to go on, no matter how devastated we all were. I think from then on we all pulled together and helped each other through it.' Two days after Timmie's death, B Watch in North Strand did a night's unpaid work for Kilbarrack's grieving A Watch as a gesture of solidarity. 'What other job would you get that in? It's something that'll never be forgotten here,' says Damian.

Profoundly moving and sad is probably the best way to describe a firefighter's funeral and the massive crowd gathering at Árd na Gréine church in Kilbarrack was a tribute to Timmie's life. 'I suppose when someone has died, there is a tendency to overstate how good a person they were. But with Timmie, it wasn't overstated, it was just right.' Timmie's other brother, Greg, who had moved to America many years before, delivered a powerful oration on the life of his younger brother, in which he spoke of his dead brother's pride at being part of DFB:

'Timmie was very proud to be a member of Dublin Fire Brigade. Their presence here today and their very supportive approach to Noeleen and our family in helping us through this tragedy is remarkable and greatly appreciated. We would like to single out Tommy Ellis, Timmie's close friend, and thank him especially. It is truly a tough time for him personally yet his entire focus is on helping Noeleen through this time. I would share with you all that Timmie is present here in the ranks today. Where he would want to be, his words being: "now lads, make sure you take care of each other."

'Timmie was very proud to be one of the lads, his silent leadership earned him great respect on the job and made him a great firefighter. I have been hearing this for the last ten years from many of his colleagues. While many of you have told the family of how great this loss is for all of you on a personal basis, it is also a great loss to the Dublin Fire Brigade. We can only ask that you honour Timmie and his memory in your day-to-day activities by maintaining your strong work ethic on the job. Timmie would not have wanted to have any of you look on the dark side of this tragedy after today, he would want you to continue to maintain the very high level of service to the citizens of Dublin and the surrounding county. He was proud to be a member of the Dublin Fire Brigade and he took great pride in its accomplishments.'

Of course after the farewells to a loved one, the togetherness and the tears, the terrible mundanity of life without the person who made working and living that much more pleasurable must be faced again. For his close friends within DFB, it was tough without Timmie. 'Listen, death is tough on everyone that's left behind,' says Tommy Ellis. 'There is nothing within Dublin Fire Brigade that makes it different for us, than it would be for anyone else, I missed my friend and I knew the other guys in the station were the same. It was a tough time. It was the most traumatic experience, with the exception of the Dublin bombings, of all my years in the Brigade.'

A Watch in Kilbarrack station would be different from the time that Timmie was killed in that emergency callout. Its two long symmetrical corridors were loud with the sound of silence. 'After Timmie, our watch as we knew it was changed, it would never be the same again,' says Tommy Ellis. 'You have to understand how important a Watch structure is to the life of a firefighter to understand how terrible it is to see it fall apart: the heart of the watch was gone,' says Damian Guilfoyle.

Today, on a late-November Sunday afternoon, only two men who served in the close brotherhood which was A Watch Number 5 are left. 'I won't be here next year, I need a break,' Damian says. Tommy Ellis says that after Timmie was gone, there was an irretrievable break in the Watch which would always be unfixable. 'We continued to be friends and to work together but in some ways, when you think about it, maybe Timmie was the glue that kept us so close during those precious years when we ate and breathed each other's lives, we were that close.' Next year, when Damian Guilfoyle leaves, the last part of that legacy will be gone.

But there is a fitting memorial to better years. Outside Kilbarrack station, a modest plaque of polished granite bears testament to Timmie's life, inaugurated on the tenth anniversary of his passing by his colleagues on A Watch. 'It occurred to me, and to the rest of the lads, that we needed something permanent, something lasting. This is our way of doing it,' says Damian. Inside Kilbarrack station a picture of A Watch in 1993 ages slowly on the wall, the smiles of a broken group of men increasingly unrecognisable to the new generations who serve in this proud station.

Remembering a Friend: The Death of Firefighter Brian Dempsey

On the evening of Saturday, 12 June 2004, the modest concrete environs of Dolphin's Barn fire station were busier then usual. People filed dutifully past the muster racks of firefighters' apparel and through the sparse corridors of the station to a small courtyard to the rear of the station. The compact space filled quickly with men and women from both within and without the service. On the red brick wall of the courtyard a black marble plaque was set, it simply read: 'In memory of FF Brian Dempsey, 1971–2003, erected by the officers and firefighters of Dolphin's Barn Fire Station.'

Fr Sean McArdle addressed the gathering at this, the official opening of the courtyard garden, a small simple space decorated with shrubs and a rockery. Many of the women wore sunglasses and many of the men, most members of Dublin Fire Brigade, coughed awkwardly into fists clenched over their mouths, feet shifting in the gravel underfoot, heads bowed, hiding tears. It was just under one year since Firefighter Brian Dempsey had passed away and his family, friends and former colleagues had gathered to remember him in this garden dedicated to his memory.

The death of Brian Dempsey had a tremendous impact on Dublin Fire Brigade. It cut through the summer of 2003 with raw emotion and desperate sadness. The funeral cortège which wound its way through the narrow streets of Donore Avenue that June was flanked by scores of stoic, stone-faced firefighters saying farewell to one of their own. Behind them a lone fire appliance carried Brian Dempsey's body. The steady beat of slow footfalls echoed softly as the honour guard, clad immaculately in Number 1 Dress Uniform, trod a slow path to St Theresa's Church. Saying goodbye to a member of the fire service in this fashion is a time-honoured tradition, passed through the ranks and through generations but the sadness on this June day was of a different, deeper kind. These firefighters were saying goodbye to a man whose time had not yet come. At the age of 31 Brian Dempsey's life had been snatched away in a tragic accident.

When Brian entered Dublin Fire Brigade in 1995, he followed in the footsteps of his father, Jimmy Dempsey, a succession between generations which constitutes an almost sacred bond in firefighting circles. Adrian Greville, a floating s/o at the time, remembers the impact Brian Dempsey had on Rathfarnham station. 'Brian had a wonderful, bubbly personality and infectious enthusiasm. That's how I best remember him. He was also a very eager and interested recruit who loved his city and who loved Dublin Fire Brigade. I think that stemmed from him being a firefighter's son, a very special relationship.' Brian was very eager to learn from men who went before him. 'He also struck up a very special friendship with the late s/o, Willy Daffy and learned a lot from him. Older members of the service love to see that in a new recruit.'

For Alan Doyle, whose father had also served within the ranks of DFB and was a friend of Jimmy Dempsey, his first meeting with Brian in Tara Street station was the start of a strong and close friendship. 'I remember meeting Brian for the first time and we started talking about music, which was a passion for the two of us. We just hit it off straight away. I remember he said to me that he would get some blank tapes, which I needed at the time. Of course, people say things like that all the time but never actually do what they say. With Brian it was different, a few days later there was a box of tapes left in my locker. That's the sort of guy he was, he just did what he said he would do.'

In late 1996, Brian was transferred to Dolphin's Barn station, one of the city's busiest. Situated on the banks of the Grand Canal, the station is in close proximity to some of the capital's toughest neighbourhoods and the firefighters in the 'Barn are a hard, seasoned breed. The station has one of the highest callout ratios for both fire and ambulance tenders and it was here that Brian Dempsey began to earn his stripes as a professional firefighter, respected and liked by those around him. He also picked up the nickname 'Bones', attributed to him by close friends Lorcan Potts and Alan Doyle. Having a nickname is *de rigeur* in Dublin Fire Brigade, they are almost a term of acceptance, so now Brian Dempsey, aka 'Bones', could stand shoulder-to-shoulder with fellow nicknamed firefighters at the Barn such as 'Bobbsy' and 'The Gobbler'.

A pragmatic bunch, the men of A Watch quickly took to their new workmate and it was at the Barn that Brian once again worked under the watchful eye of Adrian Greville. He also struck up many close friendships, in particular with Lorcan Potts. 'Brian was the sort of guy who you would like to have alongside you on a job, no matter what it was, you could rely on him,' recalls Lorcan. 'He was a very calm, controlled and balanced individual who was also very thorough.'

In the fast, hard and often cruel world of Dublin's south inner city, it is often firefighters from Dolphin's Barn station who are called upon to pick up the pieces when things go wrong. 'I remember a few tough callouts with Brian Dempsey,' said Adrian Greville. 'We saw some distressing things, nasty RTAs with tragic results, for example, and when we got back to the station we'd put the kettle on, sit down and talk about it. It was always great to have Brian at those sort of discussions.' For Alan Doyle, who served on B Watch, his friendship with Brian continued. A Watch and B Watch overlap, so as A are coming off duty, B are coming on. 'I would often go in a bit earlier so we could have a chat when he finished or vice versa. On our days off we would often meet up in the Square in Tallaght and have a plate of lasagne and chips and a chat. That was one of Brian's big passions, his food. Though of course you wouldn't think it with the nickname he had.'

When the roof of The Patriot pub in Kilmainham caught fire on 28 May 2003, A Watch from the Barn responded to the call. It was to be the last firecall for Brian Dempsey. A house next door to the pub had been having some torch-on felt applied when an accident occurred, the roof of the house caught fire and the flames began to spread: by the time Dublin Fire Brigade arrived, the flames were licking across the roof of the old pub. 'It wasn't an exceptionally bad callout,' recalls Adrian Greville, 'but the fire was beginning to take hold quite quickly, it had quite serious potential and we really needed the two pumps we had on-scene.'

The only way to tackle a roof fire is to get in close on it, strip back the roof covering and hit it head on. It's hard, physical and dangerous work. In the words of Adrian Greville, Brian was 'right in the thick of it, doing what he loved to do, fighting fire.' Then something serious happened. 'I remember seeing Brian twist awkwardly,' says Lorcan Potts. 'He was pulling something from the roof of the pub and he'd obviously done himself some serious damage, though of course that wasn't immediately apparent at the time.' Amidst the sweat and adrenaline of the scene Brian continued to work, oblivious to the internal injury that was eventually to claim his life. 'That was always Brian's way, he just got on with it,' adds Lorcan.

When the fire had been successfully extinguished, A Watch returned to the Barn. 'It was a normal afternoon, we'd just finished a firecall and were on our way back to the station for lunch and a chat,' remembers Adrian Greville. Brian Dempsey was doing the same, joining in the banter in the fire engine which always follows a job well done. That afternoon, after lunch, he got on with the job as normal, showing some new recruits some Swiftwater Rescue Training. Aimed at improving the brigade's responses to retrieval and recovery in dangerous water, it was another aspect of the job that Brian loved and he was at the forefront of propagating it throughout the fire service. The rest of that day continued normally and Brian finished his shift at 6 pm and left Dolphin's Barn station to return to his home on Donore Avenue. Nobody suspected that it would be the last shift he ever finished.

On the night of 28 May, Brian took a turn for the worse; it was obvious that whatever had happened on that firecall was far graver then initially thought. The 31-year-old was taken to hospital. The following days matured swiftly and terribly for everyone who knew Brian Dempsey as a son, a husband and of course as a firefighter. At the time, the gravity of the situation had not really taken hold. Some of his colleagues on hearing the news paid him a visit, and the banter which flowed so easily in the Barn continued at Brian's bedside. 'That was Brian, always up for a laugh. Sadly that day was one of the last times we ever talked to him,' remembers Adrian.

It is known that he died of complications arising from internal injuries he sustained on his last callout. After a first operation on his injuries, things got a lot worse. For Brian's family and close friends such as Alan Doyle, the hot days of summer 2003 turned into a nightmare. 'I remember that I wasn't concerned at all when he went into hospital. In fact, I was one of the people who urged him to go in as he was obviously in a lot of pain following the incident. I would pop in to see him and prior to his first operation, just after the June bank holiday weekend, we were talking about what we would do when he got out. In my mind it was all fairly routine but I could see that Brian was concerned about it. I remember that night, just before I left, I gave him a hug. It never occurred to me that I would never really talk to him again.'

Brian Dempsey's condition deteriorated rapidly following his operation and the prognosis was grim. He didn't give up easily, clinging to life for almost a month, his beloved wife Orla and his family keeping an almost-constant vigil at his bedside. Of course, his many colleagues from the fire service, both in Dublin and further afield, also paid regular visits. For the men of 'A' Watch at Dolphin's Barn, Brian was never far from anyone's thoughts, but the 999 calls would still continue and another firefighter sat in Brian Dempsey's seat when the sirens wailed and assistance was needed. 'We never really accepted that Brian was going to die,' says Adrian. 'We knew what the prognosis was, and it was bad, but somehow we always believed that he would get better. We got on with our jobs, that's what he would have wanted.'

For Alan, plans which he and Brian had discussed for the future began to unravel in the most tragic circumstances. 'It's amazing what you take for granted in life. I was engaged to be married and Brian had been married just under a year. When we went away for a weekend to Liverpool, I remember talking about what the future would be like. Orla, his wife, was expecting a child, I had a five-month-old son myself. We talked about joint trips away with our families. I remember he said to me, "That's what life is all about, isn't it."'

On 24 June 2003, at approximately 9 pm, Brian Dempsey died. He had never recovered full consciousness after the first operation. His wife and his family were with him at the end. It was another hot June day, and Alan Doyle remembers that it was only that day when he realised that his best friend was not going to pull through. 'There had been a lot of ups and downs throughout Brian's illness, some days when I heard he would pull through, others when he wouldn't. It was only that day, when I was waiting outside the hospital that I finally realised that this was a battle he wasn't going to win. Orla came over to me and said that it was only a matter of hours, it really hit me then.'

For Adrian Greville, it was a time he would never forget. 'I wasn't Brian's best friend, there were others who knew him a lot better then me,' said Adrian. 'However, the morning after Brian died it was my duty as Station Officer to tell the men on A Watch that our friend and colleague was dead. It was one of the hardest things I've ever had to do. I never saw grown men get so upset, crying openly and I remember finding it hard to keep my own voice from breaking. It was a terrible day.' The news of Brian's death spread rapidly throughout the Brigade, bringing with it a veil of sadness that descended over the entire organisation. To lose someone so young, friendly and popular was a devastating blow for the Barn and for Dublin Fire Brigade. 'I know every single one of us felt the same way about him,' said Lorcan. 'It was just the terrible realisation that we would never see him again.'

For Alan Doyle, a tough two years was only just beginning: 'For the men who worked with Brian, everyone understood their loss, he was a colleague and, for many, a friend. I didn't work with on A Watch with Brian, so I don't think many people understood what it was like to walk into Dolphin's Barn station and not see him at the end or the beginning of a shift. It was part of my life for so long, that I found it very hard to come to terms with. I know that death is part and parcel of life and that everyone has to deal with the loss of someone close, someone special. But for me, and other people who knew Brian, we just weren't prepared for it and that's why I think it hit us so hard. Some knew how hard it was for us, others didn't seem to understand or want to understand, but that's the way working life can be, particularly in the fire brigade.' For a career firefighter like Jimmy Dempsey, Brian's father, who served so long and so successfully in the brigade, to see his son taken from him so early in his life and his career remains incredibly difficult.

Brian Dempsey's death was not the work of fire itself, but the injuries he sustained occurred while fighting fire, while doing his job. The incident is the subject of an official Dublin Fire Brigade investigation. Meanwhile, in the days after his death, his former colleagues supported Brian's traumatised immediate family as best they could, by saying goodbye to him the firefighter's way. Personnel from Dolphin's Barn formed the honour guard that transported Brian's coffin from the funeral home to St Theresa's Church. When a traditionally cloistered organisation such as Dublin Fire Brigade emerges to say goodbye in the way which it did to honour Brian Dempsey's passing, you cannot fail to be impressed and moved if you are privileged enough to witness it.

'The funeral itself was one of the hardest things myself and the other lads have ever had to do,' recalls Adrian. 'However the support from all our firefighting colleagues was truly remarkable.' The numbers of people who gathered to say goodbye were also remarkable. 'At Mount Jerome cemetery, where Brian was finally laid to rest, I remember looking around and I couldn't believe the amount of people who had shown up to pay their final respects. It was a fitting tribute to a wonderful colleague and wonderful man.'

For Alan Doyle, the funeral was incredibly difficult, but it also made him incredibly proud. 'I remember that day was difficult, but it was also wonderful in a strange way. It proved what he meant to so many people and in many ways it validated what I had always thought of him. In traditional fashion we had a wake for him, an amazing turnout at that, there were smiles and there were tears but it was a wonderful way to say goodbye.'

While Brian's family had to come to terms with their terrible loss, the crew on A Watch in Dolphin's Barn had to get on with the job. As summer that year changed to autumn, life began to move on and firefighters in the Barn came up with the idea of a small memorial garden in the station as a permanent tribute to the man they missed so much. Over the next ten months, the garden became a focus point to alleviate the sadness that Brian's former colleagues still felt. 'It was something that we started thinking about after the funeral and then one Saturday we all just decided to pitch in and get working,' remembers Bob Murphy, another close friend of Brian Dempsey. 'There are two reasons for it really. It is intended both as a way of remembering Brian and also a place of quiet reflection where you can come to think after a tough day. It's a way of helping us move on and we've found that it has a very calming influence on us.'

When the garden was officially opened in 2003, memories of Brian Dempsey came flooding back to the surface for many at the Barn. Lorcan Potts read a tribute to his friend while Brian's sister-in-law, Valerie, also made a speech on behalf of his widow, Orla, who was still too devastated by the loss of her husband to go to the place where he loved working so much. Brian's father, Jimmy Dempsey, was also there, looking at the plaque bearing a picture of his son, a firefighter, following in the footsteps of a firefighting father. Letters and tributes were also published in *Firecall* magazine, the brigade's own members' publication, but as Station Officer Greg O'Dwyer, consultant editor of the magazine, said in his foreword to the edition commemorating Brian Dempsey's life: 'Paper and ink just won't cut it when describing Brian's sad passing. It's only our constant memories and conversations that will pay him the tribute he deserves and compensate our loss with the smiles that remembering him brings to each and every one of us.' For Alan Doyle and Brian's family and other close friends, the pain is still there. 'Without a doubt, I'm a different person as a result of knowing Brian Dempsey,' says Alan Doyle. 'I don't want it to sound trite, Brian was a normal guy, no saint just like the rest of us. For me though, he was my best friend and I think that losing someone like that from your life does have a strong and lasting impact. I'm moving on slowly, just as he would want me too, but he was my friend and I will always miss him.'

The following poem was published in *Firecall* magazine following Brian Dempsey's death, written by Firefighter Paul Marsh, A Watch, Tallaght:

Found my No. 1s in the wardrobe,
Haven't worn them since my passing-out day,
Didn't think I'd need them again,
Didn't think my friend would pass away.

Don't have to make phone calls for this one,
Everyone knows the time and place,
We'll do our crying at home,
And show up early with a brave face.

Hundreds of us gather,
So many friends in a short 31 years,
A flood of blue uniforms,
Holding back a flood of tears.

We all see death regularly,
On resusc tables and in burnt-out homes,
But this time it's different,
He's one of our own.

Sometimes memories vanish over time,
Like a breeze that can blow out a fire,
But when that flame is strong, the breeze rekindles it,
And it burns even higher.

Whether you knew him as 'Bones',
Or Brian, or even Jimmy's son,
He was a gentleman and a fireman,
And his memory will always live on.

Joe McCloskey

After surviving the horror and strife of over twenty years of conflict, the fate which befell firefighter Joe McCloskey in November 2003 is particularly sad. Stationed at Dungiven in Derry, McCloskey was a twenty-five-year veteran of the service there, displaying the kind of unsought-for dedication that is the hallmark of a successful retained service. 'We just couldn't pay the Joe McCloskeys of this world for all the extra hours they put in. In every station there are a few people like Joe, who call into their station every day, unpaid, to check that everything is ok,' CFO Colin Lammey said after Joe's death.

At approximately 1 am on the morning of Saturday, 1 November 2003, a call went through to the Dungiven Station Officer that a serious fire had been reported at the Gorteen House hotel in Limavady. Like most retained firefighters, at that hour, Joe was asleep in his bed when his pager went, summoning him to the station which was literally yards from his house. The station's appliance, with five crew, turned out to the fireground to find the hotel absolutely ablaze. The fire required the attendance of six appliances in all. All those inside the hotel had been evacuated safely but the structure itself was gutted with flame. Eager to knock down the fire as quickly as possible, the firefighters made down their hose lines and proceeded to spray massive amounts of water upon the flames.

Joe McCloskey was sent onto the roof of the hotel to spray water down upon the fire when tragedy struck. Below him, the roof had been badly scorched and as he made his way across, it gave way beneath him and the fifty-year-old plummeted into the flames below. His colleagues on the ground saw what happened and ran to his aid. Two of them were injured as they dragged Joe from the flames; he was still alive but suffering from terminal injuries. He was rushed to hospital but died the following day, Sunday 2 November at 4.15 pm.

The fire service in general and his beloved retained brigade went into deep shock, this was a terrible loss. One of Joe's fellow firefighters, Peter McDermott, told the BBC that he would miss his friend, a quiet man who regularly brought smiles to their faces: 'He was very much the heartbeat of the station, he had over twenty-five years' service and he will be extremely hard to replace. I doubt very much if we will be able to replace him. The Western Area Commander for the Northern Ireland Fire Authority, Ian Doyle added that whatever tributes were paid to Joe, they were deserved as such men were the lifeblood of the service. 'People call them part-time firefighters but really there's no such thing. His wife Marie told us that there were three people in their marriage, them and the fire service. There were many family sacrifices over the years, there was many a dinner kept warm in the oven for him. Joe lived out a childhood dream being part of the brigade and we have to take some comfort in that, and the fact that he lived all his life just a few doors away from the station.'

Before McCloskey's funeral Doyle had called to the family home and the fallen firefighter's son, 13-year-old Seamus, took pride in showing the commander his father's collection of model fire engines. Despite the massive loss which Joe's death represented for the Dungiven Brigade, Sub-Officer McDermott knew it could have been much worse. Three firefighters were injured in the dash to try and save Joe from the flames. 'I left with five crew and returned with just two.' The injured crewmembers, mostly suffering from smoke-inhalation, were attended by trained counsellors in the aftermath of the tragedy.

On Thursday 6 November, the streets of Dungiven were lined with thousands of people as the funeral cortège, two appliances and a lone piper along with his wife and five children, of Joe McCloskey made its way through the town to St Patrick's Church. There were over 500 firefighters from throughout Northern Ireland, the UK and the Republic present at the funeral. At the funeral, Andrew McCloskey told mourners that his son had died a brave and dedicated firefighter. In stations across the UK and as far away as New York, a minute's silence was held in memory of the fallen firefighter. 'It's always a shock when something like this happens, indeed it's a tragedy but the fire service goes on, Joe served the community and we will continue to serve the community as well,' said Deputy Chief Fire Officer Louis Jones.

What compounded the grief of Joe's family, friends and colleagues was that it emerged soon after, that the fire which claimed his life had been started maliciously. Later that November, two men were charged with the manslaughter of Joe McCloskey. On a Northern Ireland web forum, a man identified only as 'Gavin' summed up the feelings of many in the area: 'They deserve the highest punishment possible for tearing a good and honest family apart, for taking them to hell and back.' Dungiven firefighters bore the coffin of the father of five to its final resting place. At the graveside Seamus McCloskey, who was as fascinated with the fire service as his father, was presented with Joe McCloskey's fire helmet in poignant tribute.

Belfast Firefighters Under Fire

For Northern Ireland's firefighters the middle and the latter half of the 20th century seems to have contained little but decades of war, bombs, death and brutal conflict in which the firefighter was on the front line, with little respite. In all, between 1971 and 2003, ten firefighters lost their lives in Northern Ireland. Four of these deaths are attributable, directly or indirectly, to the conflict.

In the cities of Northern Ireland, during those years and — indeed until very recently — firefighters served their entire careers in the fire service on a war footing, stuck between warring communities who used arson and incendiary materials as daily tools in their hate-filled conflict. Although firefighters were by and large not targeted deliberately, that all changed on 6 February 1973. Before that date, only two firefighters had died, in a fire in Derry in 1971. It is widely believed that the fire in which Leading Firefighter Leonard McCartney and Andrew Wylie perished was caused as a result of terrorist actions. 6 February was a day of general strike in Northern Ireland, when the country's Loyalists were urged out to protest against the recently signed Sunningdale Agreement.

For Belfast's firefighters it would be a long day of threats, hoax calls and malicious fires. Tension was high across the city and fire crews reported seeing gangs of Loyalist gunmen active in the city's tougher areas. That evening a Catholic premises in the notorious Sandy Row area of the city was firebombed. Two fire appliances attended the incident under the command of Station Officer Jack Fell. Hose lines were being run by the firefighters from the appliances toward the burning building, one of them by Brian Douglas, Firefighter number 557, aged twenty-six. Douglas had only been in the service a short time, since 1970, and had made a good name for himself as an able and willing colleague.

That night as he ran toward a burning building, fire hose in hand, a gunman stepped forward from the shadows and opened fire with an automatic weapon. Douglas was hit several times in the abdomen and chest area, causing serious damage to major organs. He collapsed immediately onto the roadway, close to the burning building. His life drawing to a close, he was comforted by his colleagues as an ambulance was dispatched. Brian Douglas was pronounced dead on arrival at hospital. Deputy Chief Fire Officer Sidney Pollock said afterwards, 'Five of us could have been killed by that burst of fire.'

In the press of the following days, the death of Douglas did make news but not of the scale that the callousness of his murder might have expected. He was mentioned on the front page of the *Irish News*, along with the other unfortunate victims of the violence that night. Later in the paper, it was reported that a fund would be set up for his dependents by other emergency workers.

The City Fire Chief, Mr Robert Mitchell, said that nobody could understand the reason for such a terrible crime. 'We in the fire service are completely neutral,' he said. 'We are not concerned with politics or religion. It is our duty to fight fires, no matter where they are. At no time could we imagine anyone attacking us in the course of our duty.' The police were satisfied that Douglas's death was premeditated and he was not caught in any sort of crossfire as was suggested in the immediate aftermath. Mr Joseph Hall of the Fire Brigades Union said that Brian died while carrying out the finest traditions of the service, one duty with his comrades in service of the community. 'That a fireman should be killed in such circumstances has left our members and, we feel sure, all responsible members of the community aghast and angry,' said District Secretary Archie Culbert. He added that the firefighters of Northern Ireland had sought at all times, and not least during the previous difficult three-and-a-half years, to serve the people of the province impartially and often under the most trying and dangerous conditions. 'A fireman lives with danger every day by the very nature of his job. But to him there is no enemy, no "them" and "us". The only war he wages is against fire, the only victory he seeks to win is the safety of life and property. Our only appeal would be that our already hazardous task should not be made more so,' he concluded.

Allan Wright, who wrote a book on the Northern Ireland brigade in 1999, entitled *The Burning Years*, said that he worked in conjunction with Douglas but did not know him personally. He did say though that the official funeral for the firefighter was amongst the saddest he had ever seen, the cortège stretching for a quarter of a mile, at the front of which was Brian's widowed father, blind and just robbed of his only son. Speaking to the papers before the funeral he had told of his son's fascination with the fire brigade as a boy, his obsession with models of fire engines. To use Wright's words: 'I had never seen a blind man cry before. I hope it is the last time. It was so sad to see a blind man cry.'

The coffin was carried, bearing Brian's firefighter's axe and helmet, to Carnmoney cemetery on a fire appliance. Behind the black muffled bell of the appliance were two funeral hearses full of wreaths and behind them over 400 firefighters from throughout Northern Ireland, Scotland and the Republic with Dublin Fire Brigade sending a small contingent. The presence of firefighters from the Republic's capital did not go unnoticed by local media, who remembered the aid of the Southern brigades during the Blitz and said that their attending the funeral emphasised once again that, true to their international motto, 'We fight fires, not people,' firefighters know no borders or barriers.

Alongside his firefighting brothers, the funeral procession was attended by the ambulance service who assisted in carrying Brian's coffin from Whitehouse Presbyterian Church to the cemetery where he was laid to rest.

Although there was never an official inquiry into the death of Brian Douglas that night, there was some justice. Ten years later, an unemployed scaffolder from the Sandy Row area was charged with the murder of firefighter Douglas. But the dangers for firefighters did not abate as the conflict entered its darkest years.

On 16 November 1978, a car drove up to the gates of the Bass Charrington Brewery on the Glen Road in West Belfast. With one man holding a security guard at gunpoint the driver of the car drove to the warehouse of the premises and planted a bomb. It detonated several minutes later. As was their duty, the firecrews from the nearby town of Lisburn rushed to the aid of their Belfast colleagues. Among these was Sub-Officer Wesley Orr, 53, who had been with the brigade since he was just eighteen and was now in charge of the Lisburn brigade.

When they arrived on scene a massive blaze was developing, thriving on the vast quantity of alcohol inside. After setting up their firefighting equipment, Orr and four other crewmembers prepared to enter the burning building. When the firefighters entered, a second bomb detonated, ripping through the already burning building. Orr had been at the front of his men going into the fire and he caught the full brunt of the explosion. The four other firefighters were badly but not seriously injured. Wesley Orr was rushed to Musgrave Park hospital for emergency surgery but died a short time later. It is believed that whoever was responsible for the bombing left a second device to cause further death and destruction, a common terrorist tactic.

The way in which Wesley Orr died highlighted both the senselessness of the spate of violence in the North at the time and the inherent courage with which firefighters were carrying out their job in what was effectively a conflict zone. The deceased was a married man, with three children. At his funeral several days later the Bishop of Connor, Dr Arthur Butler, called on people throughout the island of Ireland to stand firm against terrorists and to remember men like Wesley Orr. He added that the firefighter's ultimate sacrifice that day, as he struggled to save another part of the city, was an example to the public that the work of the emergency services could not be taken for granted, particularly in times of crisis: 'But for their courage and determination, the toll of death and destruction would be still more frightful.'

Orr's funeral was decorated with full fire brigade honours, his colleagues from Lisburn flanking the fire appliance which bore his coffin along the three-mile route from the church to the New Blaris graveyard. As in the case of Brian Douglas, there were also brigades from the Republic present, with personnel from Dublin and Monaghan representing the firefighters of the South. Paying tribute to the deceased the mayor of Lisburn, Mrs Elsie Kelsey, said that is was so sad that a man should have to put his life at risk in the line of duty to save lives and property. She went on to say, 'Mr Orr lived up to the very high sense of duty and traditions of the fire service in Northern Ireland. He and the members of his unit did not stop to count the cost.'

Sydney Pollock, the Deputy Chief Fire Officer added that he had died as a leader, at the front of his men going into action. 'He was on the front line doing his duty. That is where all officers should be,' he added. 'Wesley was truly dedicated in his approach to the service and to the community at large. He was a great officer and a man we can ill afford to lose.' Posthumously, Orr was awarded the British Empire Medal for distinguished service. Today, he is remembered forever in a stained glass window in Lisburn Methodist church. The window was dedicated to Orr in 1985 and also to the memory of his two brothers, not in the fire service, who died the same year and his sister who had died four years earlier. Dr Hedley Plunkett, the former President of the Methodist Church, said at the ceremony at which the window was unveiled: 'The dedication (of the window) highlights the debt which the community owes to its firemen, a group of unsung heroes who constantly risk life and limb on behalf of the beleaguered people of this province, especially during the last sixteen years of wanton destruction. It is appropriate that we should memorialise one of this remarkable group of public-spirited servants who lost his life at the post of duty.'

Shootings, burnings and beatings were part and parcel of the daily dangers for firefighters in the Northern capital throughout the Troubles. Another example occurred in August 1969, before firefighters had become accustomed to the savagery of the unrest, when fire officers Billy Whyte and Dickie Sefton went to observe riots on the Lower Falls Road. Both men were seasoned firefighters and were travelling in a marked fire brigade car when they came across signs of civil unrest on the streets near the Lower Falls. The next moment, a blur of movement and the windscreen was smashed with a lead pipe. Then two petrol bombs were thrown into the car and it erupted in flames. The car spun out of control, crushing a rioter and finally coming to a stop against a wall on Peel Street, the doors bursting open on impact. Whyte was thrown through the windscreen and onto the street in flames while Sefton writhed in agony on the road trying to extinguish the fire that had engulfed his head, he was unable to breathe and running short of oxygen. A nearby lady tried to help before being chased away by rioters who then realised they had targeted firefighters instead of police.

Fire crews nearby had witnessed the brutal attack and ran to the aid of their comrades. They forced the crowds back and extinguished the flames, still swathing Whyte and Sefton who were soaked in petrol. Both men suffered horrific injuries, with particularly severe burns to their hands and heads which left them badly scarred for the remainder of their lives. Dickie Sefton's injuries were the more extreme and he spent over four months, much of it in terrible pain, in hospital before being allowed to spend Christmas with his family. He then spent three further months undergoing intense medical care. Amazingly he returned to duty in April 1970 but over the next two years he underwent seventeen often excruciating operations on his face and hands.

The attack on Sefton and Whyte was a brutal reminder to firefighters of the terrifying era into which they were but entering the early days. It shocked the entire brigade and left many questions. Would this now be considered normal? It did represent a watershed in how the Brigade conducted its operations. As Wright says in *The Burning Years*, 'Although we served all of the community, we now knew that we were safe from none.'

Conclusion

So, why do they do it?

Irish firefighters certainly don't work for the pay, the conditions or the prestige. While the pay and conditions have improved immeasurably over the past two decades, any private-sector job which comes with such a level of risk and such a demand in terms of commitment and discipline would certainly be a lot more lucrative.

As for the prestige, Irish firefighters don't seek it and they probably wouldn't get it anyway. The work of a firefighter often only makes the news as the net result of a death or serious injury: that's the nature of the media. Such an attitude is commonplace throughout much of Europe, in stark contrast to the US, where firefighters are often put on a pedestal. This is only partly due to the events of September 11 2001, but more as a result of the robust PR policy advocated by many American fire services. That's not the case here. Many do it because they feel, quite simply, that they would be good at it. Others are drawn towards it from other elements of the uniformed services, especially the army. For many it is in the blood, following in the footsteps of generations.

Most Irish firefighters whom I met during the course of compiling this book dreaded the attention in many ways. I was often pointed towards someone else who would return the compliment and lead me off in a different direction. However, despite initial hesitation, almost everyone had stories and, it would come as no surprise to firefighters, many of them were not fit to print. In fact, firefighters seem particularly content with the perception of the service as cloistered, particularly the full-time brigades. Self-promotion has very little place in the fire service, in my perception of it, certainly. I'm sure it does exist, but I'm equally sure it's very rare. Mention the dreaded word 'hero' and most often you'll get a shrug of the shoulders, a shake of the head and be told that, 'No, it's just the job.' In fact, many of the callouts are terrifyingly mundane, false alarms, bin fires or other examples of domestic stupidity, but when the bell goes, the response is the same. Up, out, gone in seconds. When the bell goes, every call is equally important.

For many brigades, again, particularly the full-time ones, but indeed brigades in general, this is a time of change. Many of the older generation are retiring, people who you will hear today's firefighters refer to as 'real firemen'. These were people who faced down terrifying infernos and situations with equipment which would often not even be contemplated in today's safety-conscious workplace. It's a changing of the guard in many respects. The passing of the older guard goes almost unnoticed of course, the way they would want it. They discuss what they have done and what they have seen with their fellow firefighters; as they say constantly, 'most of what I have seen is just for other firefighters to hear.' In place of this experience comes the new generation, far better equipped, trained and resourced than their elders to fight flame and to affect rescues in ever more extreme situations. The excesses of the modern world have created an ever-changing array of cases for the modern fire and rescue professional, from the lethal to the bizarre and the downright stupid.

It's a job that nobody contemplates doing half-heartedly, it's not fair to the public which they serve nor to the colleagues they work with every day. It's a job which demands an almost vocational allegiance, a dedication to the side of life which passes the vast majority of people by. We hear the blue light, we hear the sirens, sometimes we wonder why for a moment, then we walk away. That they stand always ready to answer the call the day that we do need them; to put themselves in harm's way for our sakes, is something for which we should always be thankful.

Bibliography

The Job
— *Review of Fire Safety and Fire Services in Ireland*; Farrell Grant Sparks Consultants; Commissioned by the Department of the Environment, Heritage and Local Government (2002).
— Information from Alan Finn, webmaster of *www.irishfireservices.com*.

Station Life
— Quotes from Gary Hayden and Terry O'Connor; *Firecall* Magazine, Volume 6, Issue 3.
— Quotes from Gerry Bell, Ray Leavy and Tony Daly; *Firecall* Magazine, Volume 6, Issue 1.
— Quotes from Nobby Clarke; *Firecall* Magazine, Volume 6, Issue 1.
— Quotes referring to Marine Emergency Response Unit; *Firecall* Magazine, Volume 7, Issue 4.
— Material in relation to stress management; *Firecall* Magazine, Volume 3, Issue 2.
— Quotes from Shay Rowe in relation to water rescue; *Firecall* Magazine, Volume 4, Issue 5.

Fire and Human Behaviour
— Quotes from Justice McCollum on Arson Attack; *Firecall* Magazine, Volume 2, Issue 1.
— Quotes from Jamie Novak on Fire Behaviour; *Report of Fire Investigators Association of Ireland*, 2005.
— Report of 3rd International Symposium on Human Behaviour in Fire, Belfast, 2004.
— Dublin Fire Brigade Training Centre, Fire Behaviour Course Student Notes.
— Quotes from Noel Kelly and Michael O'Loughlin; *Firecall* Magazine, Volume 2, Issue 4.

Famous Fires
Edgeworthstown
— *The Irish Times*, 22/24 April 1995.

— *Irish Independent*, 22 April 1995.

— *Longford Leader*, 24/28 April 1995.

Cow Comforts

— *The Irish Times*, 3 April 1999.

— Mayo County Council Official Report on Fire at Cow Comforts, April 1999.

Ballingarry Mine

— *The Irish Times*, 7 February 1973.

— *Irish Independent*, 7 February 1973.

— *Tipperary Star*, February 1973.

Halal Meats

— *Official Report of Roscommon Fire Service into the Fire at Halal Meats Factory*, January 1972.

Trinity College

— *The Irish Times*, 14 July 1984.

— *Irish Press/Evening Press*, 14 July 1984.

— *Firecall* Magazine, Volume 2, Issue 1.

Powerscourt House

— *The Irish Times*, 5/8 October 1974.

— *Irish Press/Evening Press*, 5 October 1974.

— *Irish Independent*, 5 October 1974.

— *Firecall* Magazine, Volume 4, Issue 3.

Slane Castle

— *The Irish Times*, 19/20 November 1991.

— *Irish Independent*, 19/20 November 1991.

— *Irish Press/Evening Press*, 20 November 1991.

— *Firecall* Magazine, Volume 3, Issue 2.

The Irish Times

— *The Dublin Fire Brigade: A History of The Brigade, The Fires and The Emergencies*, Tom Geraghty and Trevor Whitehead; Dublin City Council, 2004. ISBN 0946841705.

— *Firecall* Magazine, Volume 3, Issue 2.

Belfast Blitz

— *The Dublin Fire Brigade*, op. cit.

— *Firecall* Magazine, Volume 3, Issue 4.

— Archives of Drogheda Town Council, 1941.

— Archives of Dundalk Town Council, 1941.

— Material from Dun Laoghaire Historical Society, references to the *Irish News* and *Belfast Telegraph*, and reports and material relating to Hermann Goertz sourced from *Belfast Is Burning 1941: The Story of the Assistance Given by the Emergency Services from Eire following the German Bombing of Belfast*. Sean Redmond, IMPACT, 2002.

Whiddy Island

— *The Irish Times, Irish Independent, The Examiner,* 9/13 January 1979.

— Disaster at Whiddy Island, Bantry, Co. Cork: *Report of the Tribunal of Inquiry Pursuant to Resolutions Passed by Dáil Eireann on 6 March 1979 and Seanad Eireann on 8 March 1979*, Government Stationery Office, 1980.

— *Firecall* Magazine, Volume 2, Issue 1.

Calor Kosangas Explosion

— *Irish Independent*, 7 October 1975.

— *Evening Press*, 6 October 1974.

Stardust

— *Report of the Tribunal of Inquiry into the Fire at the Stardust, Artane, Dublin on 14 February, 1981*, Government Stationery Office, 1982.

— *Report of the Stardust Victims Compensation Tribunal*, Government Stationery Office, 1991.

— *The Dublin Fire Brigade*, op. cit.

Noyeks

— *Report of Dublin Fire Brigade Chief Fire Officer Thomas P. O'Brien into Noyeks Fire,* Dublin Corporation, August 1972.

— *The Irish Times*, 28 April 1972.

— *Irish Independent*, 28 April 1972.

— *Evening Press*, 28 April 1972.

— *The Dublin Fire Brigade*, op. cit.

Dublin and Monaghan Bombings
— *The Dublin Fire Brigade*, op. cit.
— *The Irish Times*, 18/19 May 1974.
— *Irish Independent*, 18/19 May 1974.
— *Irish Press*, 18 May 1974.
— *The Examiner*, 18 May 1974.
— *Sunday World*, 19 May 1974.
— *Northern Standard*, 24 May 1974.
— *Evening Press*, 31 May 1974.
— *Final Report of the Independent Commission of Inquiry into the Dublin and Monaghan Bombings.* Joint committee on Justice, Equality, Defence and Women's Rights, March 2004.

Central Hotel
— *Donegal Democrat*, 10/17 August 1980.

Charleville and Buttevant
— *Firecall* Magazine, Volume 2, Issue 2.
— *Irish Press*, 22 August 1983.
— *The Irish Times*, 22 August 1983.
— *Irish Independent*, 22 August 1983.

Raglan House and Dolphin House
— *Report on the Investigation into the Explosion at Raglan House, Ballsbridge, Dublin 4, 1 January 1987*; Compiled by Cremer & Warner for the Minister for Energy. Government Stationery Office, 1987.
— *The Irish Times*, 2 January 1987.
— *Irish Independent*, 2 January 1987.
— *Irish Press*, 2 January 1987.

Saying Goodbye
Willie Bermingham
— *The Dublin Fire Brigade*, op. cit.
— *Alone Again*, Willie Bermingham and Liam O'Cuanaigh, ALONE 1983.
— *Brigade Call* Magazine, Christmas 1977.
— *Brigade Call* Magazine, Summer 1991.

Brian Dempsey
— *Firecall* Magazine, Volume 5, Issue 3; Volume 6, Issue 1.

Northern Ireland

— *Irish News*, 7 February 1973.
— *Burning Issues: A Fire Officer's Memoirs of Belfast's Firefighters Baptism of Terrorism During the Troubled Years 1968–1988*; Allan Wright, Rosepark Publishing 1999.
— *The Flaming Truth: A History of the Belfast Fire Brigade*, William Broadhurst and Henry Welsh, Flaming, 2001. ISBN 0954159705.